Advanced SPICE Model for GaN HEMTs (ASM-HEMT)

Sourabh Khandelwal

Advanced SPICE Model for GaN HEMTs (ASM-HEMT)

A New Industry-Standard Compact Model for GaN-based Power and RF Circuit Design

Springer

Sourabh Khandelwal
School of Engineering
Macquarie University
Sydney, NSW, Australia

ISBN 978-3-030-77729-6 ISBN 978-3-030-77730-2 (eBook)
https://doi.org/10.1007/978-3-030-77730-2

This Springer imprint is published by the registered company Springer Nature Switzerland AG
The registered company address is: Gewerbestrasse 11, 6330 Cham, Switzerland

To all ASM-HEMT users and supporters. Your feedback helps the model to keep improving.

Preface

The idea of this book came from ASM-HEMT model users and my students. During my interactions with them, a need for more detailed text on ASM-HEMT and GaN-HEMT technology from a device modeling perspective became apparent. The students were curious about the various applications of GaN HEMT technology, and how device modeling changes from one application to the other. ASM-HEMT model users who are practicing device modeling engineers expressed the need for a more detailed text on ASM-HEMT model as compared to the technical manual. During my lectures it also became clear that a clear distinction between different styles of compact models is also a topic which needs attention. This background led to the book.

I hope that this book serves following purposes. Firstly, I hope to describe basics of GaN HEMT technology and its two diverse applications to students, engineers, and researchers new to this technology. Secondly, I aspire to educate engineers and student new to compact modeling on the various types, aspects, and challenges of compact modeling. This book can be viewed as a book describing a process to develop a physics-based compact model, with ASM-HEMT as an example. Another important purpose I hope this book serves is to inform the experienced users on the detailed derivations of various parts of the industry standard ASM-HEMT compact model. This book also includes resources which are continuously updated as new versions of ASM-HEMT model are created.

GaN HEMT is a fast-emerging semiconductor technology, and ASM-HEMT is a very important tool for circuit and product design in this technology. Overall, I hope this book informs engineers, students, and researchers working in this area about ASM-HEMT model and enable them to exploit the tremendous potentials of GaN technology in circuits and products.

Sydney, NSW, Australia
Oct 21, 2021

Sourabh Khandelwal

Acknowledgments

It has been more than 10 years since I have been focusing on research and development of ASM-HEMT model. During this time, I have been fortunate to work with some of the best minds in the area of semiconductor device modeling. I am grateful to such interactions which range from over the coffee chat to deep technical discussions in conferences. I am thankful to Prof. T. A. Fjeldly for his guidance during my PhD days where I was starting this model development. My PhD co-advisor Prof. T. Ytterdal also helped me on several aspects of model development. During this time, I also had numerous positive interactions with Prof. B. Iniguez, Dr. Rob Davis, Dr. Tariq Lodhi, and Dr. F. Mulugeta.

My research on ASM-HEMT model continued beyond my PhD research. During postdoctoral research while working on the BSIM models, I had the great fortune of interactions with Prof. Chenming Hu of the University of California Berkeley. Prof. Chenming Hu's knowledge and experience of the semiconductor device modeling continues to guide me even to this day. His knowledge is only foreshadowed by his amazingly helpful human side. I am forever indebted to him for this work. Later, during postdoc, I also interacted with Prof. S. Salahuddin. His unmatched insights into device physics have helped me and will continue to help me in numerous ways.

During postdoc years, and later through ASM-HEMT modeling research, I also got help and support from Prof. Y. S. Chauhan and his students. They include S. A. Ahsan, S. Ghosh, and A. Dasgupta, among others. I am also grateful to have several of my students and postdocs help me with the continuous development of this model. They include J. Hodges, D. Mahajan, S. A. Albaharani, H. Eslahi, and N. Reddy.

Today, as ASM-HEMT model is being used by several companies and research groups around the world, I am fortunate to interact with users on different perspectives and continuously improve this model. G. Coram has helped on numerous occasions to improve the model code. C. Shaw and S. Mertens have helped me navigate the industry standardization process of this model, and they continue to do so.

Last but not least, over the years, my wife has helped me through all the phases of life as I focus on ASM-HEMT model development. Her patience and support have what made ASM-HEMT model and this book possible.

Contents

1 Gallium Nitride Semiconductor Devices ... 1
- 1.1 Introduction ... 1
- 1.2 Application-Level View of Transistor ... 2
  - 1.2.1 Radio-Frequency Power Amplifier Application ... 2
  - 1.2.2 Power Electronics Switching Application ... 5
- 1.3 GaN Material Properties ... 6
  - 1.3.1 Band-Gap and Breakdown Electric Field ... 7
  - 1.3.2 Electron Mobility ... 7
  - 1.3.3 Saturation Velocity ... 7
- 1.4 Brief History and Key Advances in GaN Transistors ... 7
- 1.5 Conclusion ... 8

2 Compact Modeling ... 9
- 2.1 Introduction to Compact Models ... 9
- 2.2 Types of Compact Models ... 10
  - 2.2.1 Measurement-Based Compact Models ... 11
  - 2.2.2 Empirical Compact Models ... 11
  - 2.2.3 TCAD Models ... 12
  - 2.2.4 Neural Network Based Compact Models ... 13
  - 2.2.5 Physics-Based Compact Models ... 14
- 2.3 Comparison Among Different Compact Models ... 15
- 2.4 GaN Compact Models ... 17
- 2.5 Compact Model Coalition and Industry Standard GaN Compact Models ... 19

3 Introduction to ASM-HEMT Compact Model ... 21
- 3.1 User-Level View of ASM-HEMT Compact Model ... 21
  - 3.1.1 Device Geometry Parameters ... 21
  - 3.1.2 Model Configuration Switches ... 23
  - 3.1.3 Device Physics Parameters ... 27
  - 3.1.4 Smoothing Parameters ... 27

3.2 Components of ASM-HEMT Compact Model 27
3.3 ASM-HEMT Model Features 29

4 Core Formulations in ASM-HEMT Model 33
4.1 Introduction 33
4.2 Two-Dimensional Electron Gas Model 33
4.3 Surface-Potential Model 37
4.4 Terminal Charge Model 39
4.5 Core Drain Current Model 42
4.5.1 Mechanisms of Carrier Transport 43

5 Non-ideal Effects in Device Current and Their Modeling 47
5.1 Introduction 47
5.2 Pinch-Off Effect 47
5.3 Velocity Saturation Effect 48
5.4 Non-ideal Sub-threshold Slope 50
5.5 Mobility Dependence of Vertical Field 50
5.6 Channel-Length Modulation Effect 53
5.7 Drain-Induced Barrier Lowering Effect 54
5.8 Non-linear Access Region Resistance Model 55

6 Trapping Models 63
6.1 Introduction 63
6.2 Model with Symmetric Time-Constants 64
6.3 Model with Asymmetric Time-Constants 65
6.4 Trapping Model Based on Recombination Generation 66
6.5 Measurement Results 67
6.6 Circuit Implementation of a Trap Center 70
6.6.1 Rates of Charge Capture and Emission at a Trap Center 70
6.6.2 Control Potential of a Trap Center 74
6.6.3 Current Injection into a Trap-Center Region 75
6.6.4 Circuit Model of a Trap Center 76
6.6.5 Steady-State Response 77
6.6.6 Circuit Model for an Electron-Trapping Center 78
6.7 Discussion 78

7 Non-Ideal Effects in GaN Capacitances and Their Modeling 83
7.1 Introduction 83
7.2 Impact of Fringe Capacitances 84
7.3 Impact of Field-Plates on Device Capacitance 85
7.3.1 Gate–Source Capacitance 86
7.3.2 Gate–Drain Capacitance 87
7.3.3 Drain–Source Capacitance 89
7.4 Impact of High Field Region on Capacitance 90
7.5 Impact of Charge Trapping on Capacitance 95
7.6 Substrate Bias Effects and Substrate Capacitance 97

8 Gate Current Model ... 101
8.1 Diode-Based Formulation ... 101
8.2 Model Based on Frenkel–Poole Mechanism ... 102
8.3 Gate Current with Consistency with Drain Current ... 102
8.3.1 Thermionic Emission Model ... 102
8.3.2 Poole–Frenkel Model ... 104
8.3.3 Fowler–Nordheim Model ... 108

9 Effect of Ambient Temperature on GaN Device ... 115
9.1 Introduction ... 115
9.2 Temperature Dependence of Parameters Affecting Drain Current ... 116
9.2.1 Cut-Off Voltage ... 117
9.2.2 Mobility ... 118
9.2.3 Saturation Velocity ... 119
9.2.4 Access Region Charge Density ... 121
9.2.5 Contact Resistance ... 121
9.3 Temperature Dependence of Gate Current Parameters ... 122
9.4 Self-Heating Effect and Modeling for Multiple Ambient Temperatures ... 122

10 Noise Models ... 125
10.1 Flicker Noise ... 125
10.2 Thermal Noise ... 127

11 Parameter Extraction in ASM-HEMT Model ... 131
11.1 GaN RF Device Extraction Flow ... 131
11.1.1 DC Parameter Extraction ... 131
11.1.2 S-Parameter Model ... 132
11.1.3 Non-Linear Large Signal Model ... 135
11.2 GaN Power Device Extraction Flow ... 137
11.2.1 I-V Parameter Extraction ... 137
11.2.2 Capacitance Modeling ... 140
11.2.3 Modeling the Package Parasitics ... 142
11.3 Model Quality Tests ... 146
11.3.1 Gummel Symmetry Test Results ... 146
11.3.2 Harmonic Balance Simulation Results ... 148
11.3.3 AC Symmetry Test Results ... 148
11.3.4 Reciprocity Test Results ... 149

12 Advance Simulations with ASM-HEMT Model ... 151
12.1 Manufacturing Variation and RF Device Stability Analysis ... 151
12.1.1 Custom Python Script ... 153
12.1.2 Non-Linear Stability Analysis in the Presence of Manufacturing Variations ... 156
12.1.3 Nominal Device Non-Linear Stability Analysis ... 156
12.2 GaN Power Device Circuit Simulations Using ASM-HEMT ... 161

12.3 Design Technology Co-optimization Using ASM-HEMT 166
12.3.1 Barrier Thickness Dependence of AM/PM 166
12.3.2 Field-Plate Dependence of AM/PM 169

13 Resources for ASM-HEMT Model Users 173
13.1 GITHUB Repository 173

References 175

Index 181

Acronyms

AlGaN	Aluminum Gallium Nitride
ASM-HEMT	Advance SPICE Model for High Electron Mobility Transistors
CMC	Compact Model Coalition
EE HEMT	EESof Scalable Nonlinear High Electron Mobility Transistors Model
FDSOI	Fully Depleted Silicon on Insulator
FinFET	Fin-shaped Field Effect Transistor
FP	Field-Plate
GaN	Gallium Nitride
GFP	Gate-connected Field-Plate
HEMT	High Electron Mobility Transistor
HSP	High Electron Surface-Potential-based model
MOSFET	Metal Oxide Field Effect Transistor
MVSG	MIT Virtual Source GaN model
PSD	Power Spectral Density
RF	Radio Frequency
SFP	Source-connected Field-Plate
SOI	Silicon-on-Insulator
SPICE	Simulation Program with Integrated Circuit Emphasis
TCAD	Technology Computer Aided Design
UID	Unintentionally Doped
2-DEG	Two-Dimensional Electron Gas

Chapter 1
Gallium Nitride Semiconductor Devices

1.1 Introduction

Transistors are ubiquitous circuit elements. They are deployed in electronics circuits used in applications ranging from microprocessors to cell phones to power distribution electronics. These applications deploy transistors as a switch and also as signal amplifying device. As an electronic switch, the transistor should ideally provide no resistance when turned on and an infinite resistance when turned off. In reality, any transistor made from a semiconducting material will always have a finite resistance. This is due to the limited amount of charge in the channel and the finite speed of carriers flowing in the channel. Similarly, in off state the resistance of a realistic device will not be infinite but finite value due to the different leakage mechanisms in the device. As an amplifying device, the nature of the output characteristics of transistors is used to provide amplification of a small signal. Transistors with a large input-to-output transfer or the trans-conductance are used to amplify the input signal. The ability to amplify the signals again depends on the semiconductor material properties.

Silicon-based transistors have revolutionized the world of electronic gadgets for decades now. The relentless progress in silicon-based transistor performance, as dictated by Moore's law, has led to products unimaginable without the progress in silicon transistor technology. Moore's law and this progression are slowing down and coming to an end now. Additionally, for applications concerned with high power, high frequency, and high voltage handling capabilities, silicon has material limitations. Wide band-gap semiconductors such as GaN and SiC offer significant advantages in these applications. In this chapter, we discuss what makes a good transistor for power amplifier and switching application. Our focus here is on two different types of applications: 1. signal power amplification at radio frequencies and 2. high speed, high voltage switching used in power electronics. These two applications are chosen to show the reader how the properties of GaN-based transistors suit the needs to these two differing applications quite well. This

S. Khandelwal, *Advanced SPICE Model for GaN HEMTs (ASM-HEMT)*,
https://doi.org/10.1007/978-3-030-77730-2_1

also provides a view to the reader that while seemingly different, these applications have quite a bit common when comes to transistor technology suited for these applications. For these applications, we look at some key relations that define performance and efficiency. A discussion is on how change in transistor size causes different trade-offs in performance and efficiency metrics. A discussion on these is presented below.

1.2 Application-Level View of Transistor

1.2.1 Radio-Frequency Power Amplifier Application

We review in this section the desirable properties of a semiconductor device to be suited for radio-frequency (RF) power amplifier applications. This can be understood from basic power amplifier formulations. A general power amplifier configuration is shown in Fig. 1.1. The input power is amplified by the active semiconductor device. Matching networks at the input and output minimize the undesirable reflections. As the input signal swings, the current and voltage at the output terminal of the transistor cover various regions of the output characteristics of the transistor. Power being the product of current (I) and voltage (V) as shown in Eq. 1.1, it is large if the swing in current (ΔI) and voltage (ΔV) is both as large as possible. This signal swing in current (ΔI) and voltage (ΔV) can be visualized in Fig. 1.2.

$$P_{out} = \Delta I \cdot \Delta V. \tag{1.1}$$

The current and voltage swings are

$$\Delta I = I_{max} - I_{min} \tag{1.2}$$

$$\Delta V = V_{max} - V_{knee}. \tag{1.3}$$

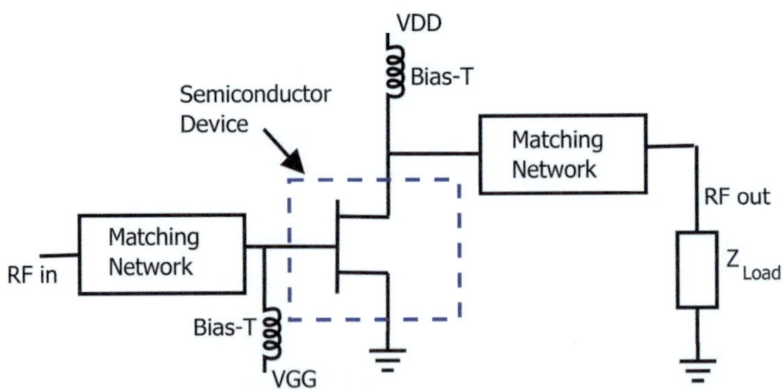

Fig. 1.1 Generic power amplifier schematic showing semiconductor device, matching networks, and load

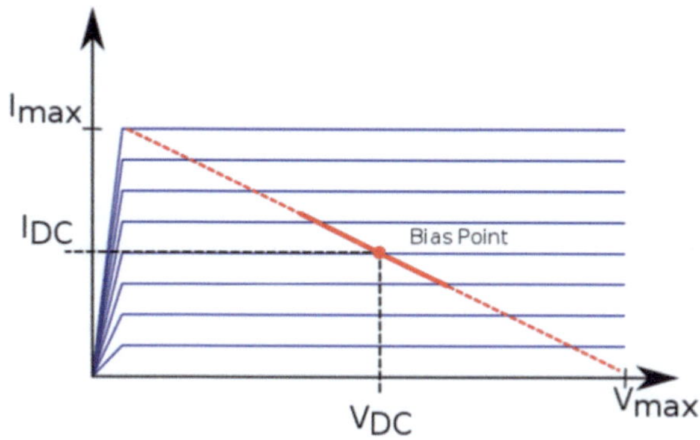

Fig. 1.2 Current-voltage plane and the load-line swing in a typical power amplifier application

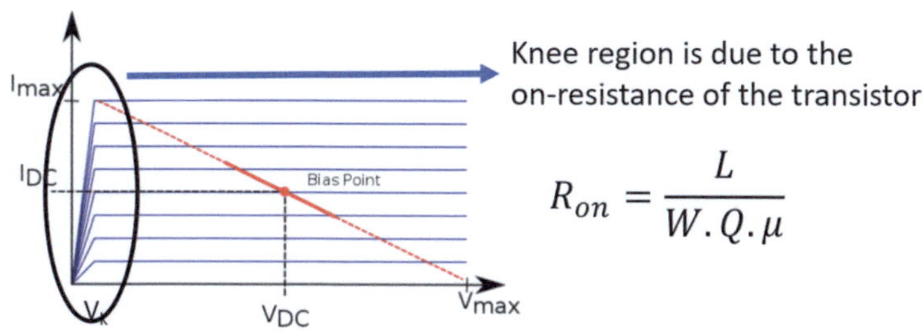

Fig. 1.3 Current-voltage relation and the on-resistance of the device. On-resistance affects the knee voltage

From Eqs. 1.1, 1.2, and 1.3, one can see that to maximize P_{out}, both ΔI and ΔV should be maximized. From Eq. 1.2, ΔI is maximum when I_{max} is large. For specific width of the device, I_max is related to the saturation velocity and the charge in the channel of the transistor, which is given by

$$I_{max} = Q_{ch} \cdot v_{sat}, \tag{1.4}$$

where I_{max} is the maximum current per unit width of the device, Q_{ch} is the channel charge per unit area, and v_{sat} is the saturation velocity. In Eq. 1.3, V_{max} and V_{knee} are related to the fundamental device properties. V_{max} is limited by the breakdown voltage of the device. For reliable operation, V_{max} should be kept below the breakdown voltage. Breakdown voltage $V_{breakdown}$ itself is related to the critical electric field E_c. E_c is related to the band-gap E_g of the semiconductor material of the device. A high E_g results in high E_c leading to a high $V_{breakdown}$ and in turn a high V_{max}. V_{knee} in Eq. 1.3 depends on the series contact resistance and mobility of the semiconductor device. V_{kneee} increases with the series resistance, while it decreases with the mobility. This is shown in Fig. 1.3.

From the above discussions, it becomes clear how the maximum output power P_{max} is related to key properties of the semiconductor material of the active device. The device properties, mobility, saturation velocity, breakdown voltage, and series resistance are critical for extracting large power from the device.

In addition to the maximum power, for radio-frequency power amplifier application, the device needs to be able to respond to the high frequency input signal. Device response should be faster than the frequency of the signal it is processing. The response of the device is also related to the semiconductor material properties. The unity-gain frequency f_t is a popular metric indicative of device speed. f_t defines at what frequency the short-circuit current gain from a transistor becomes unity. f_t can be approximated as

$$f_t = g_m/2\pi(C_{gs} + C_{gd}), \tag{1.5}$$

where g_m is the trans-conductance and C_{gs} and C_{gd} are the device capacitances. A higher f_t indicates a device with higher speed. g_m is proportional to the saturation velocity, which is the property of semiconductor material. g_m is also proportional to the width W of the device. C_{gs} and C_{gd} are proportional to the size (gate width and length) of the device. From the equations and discussion above, one may think that increasing width of the transistor should be a straightforward approach to get a large output power. Increasing width increases I_{max}, which increases P_{max}, and additionally, f_t remains the same as both g_m, C_{gs}, and C_{gd} scale up proportionally with width. Unfortunately, this does not happen in reality; in fact, both I_{max} and g_m do not scale favorably as width is increased arbitrarily. This is discussed below:

- Scaling I_{max} with width: There are two aspects of I_{max} we need to consider as we try to understand its scaling with the width. First, consider that static I_{max}. This is the case when the input voltages are applied for sufficiently large time till the current to reach its static value. In this case, a larger amount of current causes local temperature in the transistor channel to increase and go beyond the ambient temperature. An increase in local temperature causes mobility to decrease, which can lead to a decrease in the achievable current. Next, consider the case of I_{max} achievable as width is increased for the case when input signal is not static but is a radio-frequency signal. This is the more relevant case for the RF power amplifier application. To achieve I_{max} first, the input voltages at the transistor input terminals should reach to the values needed for I_{max}. However, in the case of dynamic signal, this is dependent on how fast capacitance associated with each transistor terminal can be charged. This is illustrated in Fig. 1.4. As the capacitance at the device terminals increases with the size (length and width) of the transistor, the terminal voltages cannot reach the value needed to achieve the maximum current when the signal is moving at high speed. This is the key limitation in increasing the width of the device to be too large. As the device cannot reach to input conditions of maximum current, it does not deliver high output power.

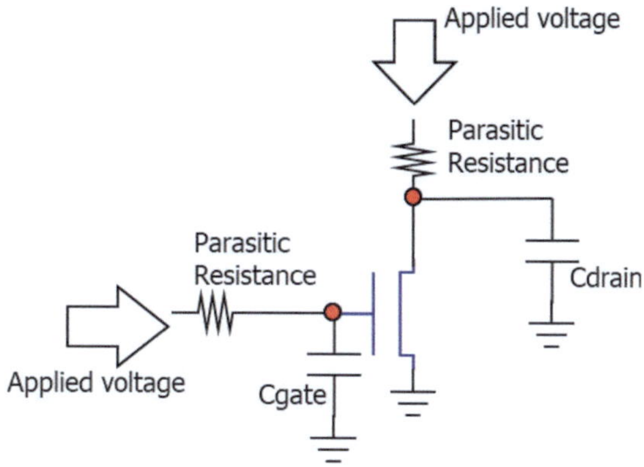

Fig. 1.4 Transistor intrinsic terminals need to be charged to the appropriate voltages to achieve the maximum current I_{max}. The charging time depends on the time-constant of parasitic resistance and the terminal capacitance

- Scaling g_m with width: The discussion about the I_{max} above applies to g_m too. While in the case of slow moving signal, the terminal input conditions can reach the needed state, in the case of fast signals charging the large terminal capacitance of a wide device becomes a limitation.

It is due to this interdependence that the transistor width cannot be arbitrarily increased for higher current and output power. An increase in width increases current, but it also increases capacitance making the transistor slower.

1.2.2 Power Electronics Switching Application

It is quite interesting to note that several device level figures of merits discussed above apply to the power device switching application too. In power electronics applications, the transistor moves from off state to on state in the switching operation. During switching activity, there are two types of losses: static and dynamic losses. Static loss is due to the finite resistance of the transistor and is given by

$$P_{static} = I_{on}^2 \cdot R_{on}, \tag{1.6}$$

where I_{on} is the current in the device and R_{on} is the resistance of the device. One can see that a lower R_{on} is desirable to have low static loss. R_{on} can be reduced by increasing the total width or size of the device. However, width cannot be increased arbitrarily due to the dynamic loss. Increasing width also limits the switching speed of the transistor as discussed above for the power amplifier case. R_{on} of the device is also dependent on the mobility of the channel material and the contact resistance of the contact made to the semiconducting channel. R_{on} is given by

$$R_{on} = L/(W \cdot \mu \cdot Q). \tag{1.7}$$

In Eq. 1.7, L is the gate length, μ is the mobility of charge carriers, and Q is the charge density per unit area under the gate contact. The on-resistance decides the static loss of the power switch, and it also decides the knee voltage for the power amplifier application. The semiconducting material in which transistor with a low knee voltage can be made is also favorable for power electronics switch.

In addition to the static loss, dynamic losses also occur in a switching device. Dynamic losses are given by

$$P_{dynamic} = C_{out} \cdot V_{ds}^2 \cdot f_{sw}, \tag{1.8}$$

where C_{out} is the total output capacitance, V_{ds} is the switching voltage, and f_{sw} is the switching frequency. C_{out} is proportional to the total width of the device. Hence, the dynamic losses increase if the size of the device is increased, provided that device can still be switched at the same speed.

The power transistors have a rated maximum current, and it depends on the width of the transistor needed to achieve this current. This puts a limit on the switching frequency, as large width transistor will have bigger terminal capacitance. As a result, a semiconducting technology that can achieve a highest current per unit width is desirable. This requirement is similar for power amplifier application too. The above discussions on power amplifier and power electronic switching point to a similar trade-off in the two applications between device resistance and size, i.e., the total device width.

1.3 GaN Material Properties

The key material properties that govern the device performance in power amplifier and power electronics application are the ones that affect the on-resistance, breakdown voltage, and maximum current of the device. As all these parameters are dependent on the ambient temperature, thermal performance of the material is also critical. This is especially important in power applications as the local device temperature can reach well above ambient temperature, while the device is in operation.

Key material parameters are compared between GaN, SiC, Si, and GaAs in Table 1.1. A discussion on each that relates these material properties to the desirable properties discussed above is presented below.

Table 1.1 Comparison of material parameters of different device technologies relevant parameters for RF and power applications

Semiconductor	Mobility (cm^2/V.s)	Saturation velocity (cm/s)	Breakdown electric field (MV/cm)	Thermal conductivity W/cm/K
GaN	1500	2.7×10^7	3.0	1.5
Silicon	1300	1.0×10^7	0.3	1.5
SiC	950	2.0×10^7	2.4	3.0
GaAs	6000	1.3×10^7	0.4	0.5

1.3.1 Band-Gap and Breakdown Electric Field

Band-gap (E_g) is related to the strength of chemical bonds in the semiconductor material. A larger E_g indicates to break the electron away from the atom one needs to supply it with larger energy. A higher band-gap translates into higher breakdown electric field (E_c) of the semiconductor. GaN with a band-gap of 3.39 eV is highest among silicon, SiC, and GaAs. The higher breakdown electric field means that for a given length of the semiconducting material, GaN can sustain highest voltage without breakdown among these materials. We have seen from earlier discussions that a higher breakdown voltage is beneficial for both power amplification and power electronic switching application. Thus, a high band-gap makes GaN an excellent choice for both these applications.

1.3.2 Electron Mobility

Electron mobility μ is related to the on-resistance and the knee voltage in power electronics and power amplifier application. A high mobility makes the on-resistance and the knee voltage lower. This reduces losses for power electronic switching as seen in Sect. 1.2.2. A higher mobility also reduced the knee voltage and increases the maximum voltage swing and, hence, the output power for power amplifier application. GaN with a good mobility in 2-DEG of 1500 cm^2/V.sec compares favorably as compared to silicon and SiC.

1.3.3 Saturation Velocity

Saturation velocity is important for achieving a large maximum current. A higher value is favorable. This is more relevant from power amplifier applications. GaN with a reasonably high saturation velocity in 2-DEG scores favorably on this parameter too.

Based on the above parameters, the combination of high electron mobility, saturation velocity, and breakdown electric field offered by GaN makes it an excellent candidate for both power amplifier and power electronic switching applications.

1.4 Brief History and Key Advances in GaN Transistors

GaN semiconductor technology development started in 1990s. Shuji Nakamura then at Nichia Chemical Industries used the material to fabricate blue, violet, and white light emitting diodes (LEDs) [1, 2]. This success in LEDs and attractive material properties directed attention and funding on development of high frequency and high power GaN transistor technology [3]. In 1991 M. A. Khan et al. reported formation for two-dimensional electron gas (2-DEG) at the AlGaN/GaN heterojunction [4]. The first GaN-based metal semiconductor FET (MESFET) and high electron mobility transistor (HEMT) were reported in 1991 [5].

A key technological advancement for GaN-based electronics has been finding the ways to grow GaN semiconducting material with low defects on foreign substrates. Researchers developed ways to grow GaN on SiC, silicon, and sapphire substrate. Nowadays, diamond as a substrate for GaN devices is also being pursued by several groups. One of the milestones toward GaN growth on foreign substrate was achieved by Amano et al. in 1986 [6]. Amano et al. showed GaN films growth at low temperature (600 deg C) by using metal organic chemical vapor deposition (MOCVD) using AlN nucleation layer [7]. Today, MOCVD is the workhorse in the industry for growth in high quality GaN films on foreign substrates.

As acceptable quality of GaN films was being repetitively being achieved by various groups with the above breakthroughs, GaN transistor design gained interest and momentum. Highly favorable material properties of GaN material for high frequency high power applications attracted several researchers [8–11]. The device performance improvements have been made possible by a series of key innovations. First, because of the high power density on GaN device, thermal dissipation is an important issue. Toward this end, thick air bridge technology with gold plating has been developed. Thick air bridge allows easy dissipation and also provides a low inductance path toward ground. To reduce the gate resistance of the GaN device working at radio frequencies, T-gate or the mushroom gate structure has been adopted [12]. T-gate allows the gate length to be short and still has low gate resistance [13]. Another important device design feature has been the field-plates [14, 15]. Although GaN material can withstand high voltages, application of high voltage without carefully designed devices can raise the electric field within the device go beyond the breakdown electric field. This is because sharp terminations and transitions within the layers of the device can have field crowding effect. Gate- and source-connected field-plates are deployed in the device to avoid any local electric field to be too large in the device. Field-plates are very commonly used in GaN technologies today.

1.5 Conclusion

This chapter presented what are the key requirements for a transistor for two different applications: 1. Power amplifier application and 2. Power electronics switching applications. From the review of needs in these applications, we established that low on-resistance, high breakdown voltage, high saturation velocity, and high channel mobility are the properties that help both the applications. After the review of the circuit level needs, we compare GaN, silicon, SiC, and GaAs material properties. With the comparison of the semiconductor material properties, we see that how GaN has favorable properties that suit these applications.

Chapter 2
Compact Modeling

2.1 Introduction to Compact Models

Circuit simulations are indispensably required in the design of today's complex circuits. Without accurate and predictive simulations, the design process can become prohibitively time-consuming and expensive for multiple design runs till the necessary performance is met. Circuits can involve one or many components such as resistors, capacitors, diodes, etc. The circuit simulations need mathematical description of terminal characteristics of all the electrical components used in the circuit to simulate its behavior. The simulators then use these mathematical descriptions within their algorithm to solve the Kirchhoff's current and voltage laws. These mathematical descriptions of electrical components' terminal behavior are known as compact models. These are also known as SPICE models, SPICE being the first circuit simulator.

A very simple example of a compact model is electrical model of a resistor. For a resistor of value R, the terminal characteristics of the resistor can be summarized by the following equation:

$$I(a, b) = V(a, b)/R. \tag{2.1}$$

In Eq. 2.1, a and b are the two terminals of the resistor R. This mathematical description of the resistor can be used by the simulator to solve the circuit in which R is deployed. This is a simple model. Now, if we want to make this model work at different ambient temperatures, we should include the ambient temperature dependence of R. Including the ambient temperature dependence will increase the model complexity to be

$$I(a, b) = V(a, b)/(R + \alpha \cdot T). \tag{2.2}$$

S. Khandelwal, *Advanced SPICE Model for GaN HEMTs (ASM-HEMT)*,
https://doi.org/10.1007/978-3-030-77730-2_2

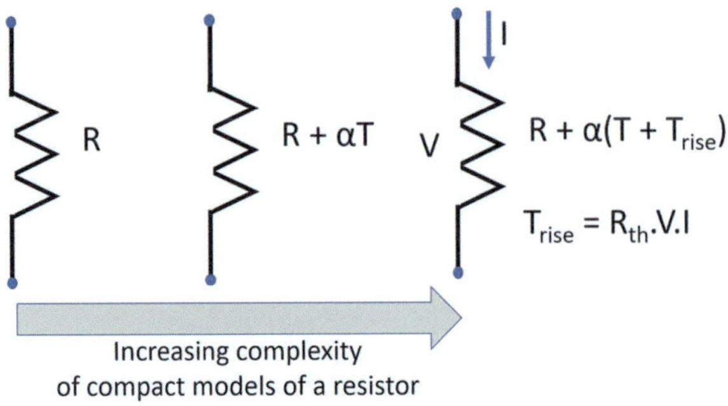

Fig. 2.1 Increasing complexity of resistor compact model as model is developed for increasing accuracy

Now, it is understood that if an increasingly large current is passed through the resistor, the resistor can get heated. As a result, the local temperature of the resistor can go above the ambient temperature. As a result, the resistor will behave non-linearly with the current flowing through it. If we want to include this effect into the compact model, we need to update the formulation further. One way of incorporating this heating effect into the model is

$$T_{rise} = I(a,b) * V(a,b) * R_{th} \tag{2.3}$$

$$I(a,b) = V(a,b)/(R + \alpha \cdot (T + T_{rise})). \tag{2.4}$$

The revised model of the resistor is now non-linear and more accurate as it includes ambient temperature and local heating effects. The formulations Eqs. 2.3 and 2.4 need to be self-consistently solved for accurate model of the resistor. One can see from the example above that depending on the need, model for a simple resistor can also become fairly complex and non-linear. This is summarized in Fig. 2.1. Nevertheless, the complexity may be absolutely required by the needs of simulations. This simple example introduces the compact models. Next, we discuss what are the different types of compact models.

2.2 Types of Compact Models

There can be many different styles of compact models of the same electrical component. These different styles of compact models vary in their development flow, capabilities, development time, and several other measures. A very detailed distinction among different types of compact models is not attempted here. A generally broad category is drawn on the types of compact models.

Measured Data Table

Drain Voltage	Drain Current
0	0
1	0.5
2	1
3	1.5
4	2
5	2.3
6	2.5
7	2.65
8	2.7
9	2.7
10	2.7

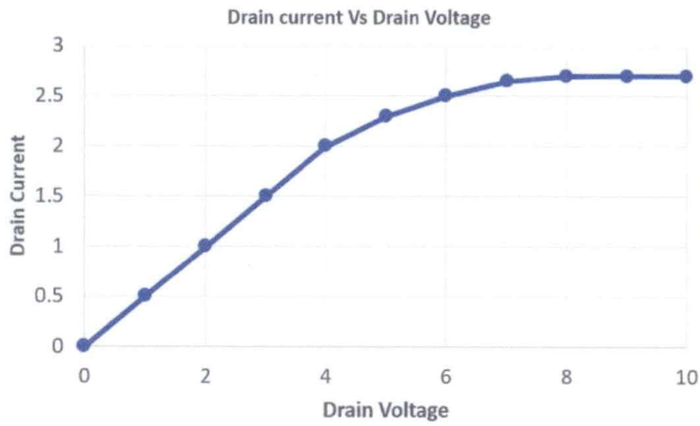

Fig. 2.2 Measurement-based model uses tables of data as a way of modeling the device behavior

2.2.1 Measurement-Based Compact Models

These compact models make use of the measured data on the device as a form of the model [16–19]. This is a simple yet effective approach for developing a compact model. In these models, the terminal characteristics of the device is characterized experimentally. The measured data is organized in the form of table identifying the input and output variables. The measurement data tables themselves serve as a compact model for the device. Almost all popular simulators allow using measured data as a way to model the device. This modeling approach is shown in Fig. 2.2.

While this approach of creating a model is simple and fast, there are some shortcomings. First, any error in the measured data becomes part of the model. Second, the measurements will be performed for certain minimum step size of the input variation and also a maximum input value. This may be related to measurement equipment, time available to take the measurement, etc. As a result, for points not included in the measurement data, the simulator has to use interpolation or extrapolation. This can lead to serious errors if the device is non-linear. Third, usually the measurement data is typically not smooth if higher order derivatives are taken. However, simulator uses higher order derivatives to solve the circuit. This may lead to non-physical results or non-convergence. Nevertheless, measurement-based compact models are widely used for their ease of development.

2.2.2 Empirical Compact Models

Empirical compact models are mathematical descriptions of terminal characteristics of the device derived by simply fitting the measured data with appropriate mathematical functions [20–23]. These models can be created from just one data-set on a device or they may also be created by looking at multiple devices and developing mathematical formulations that describe major trends in the data. An example is shown in Fig. 2.3.

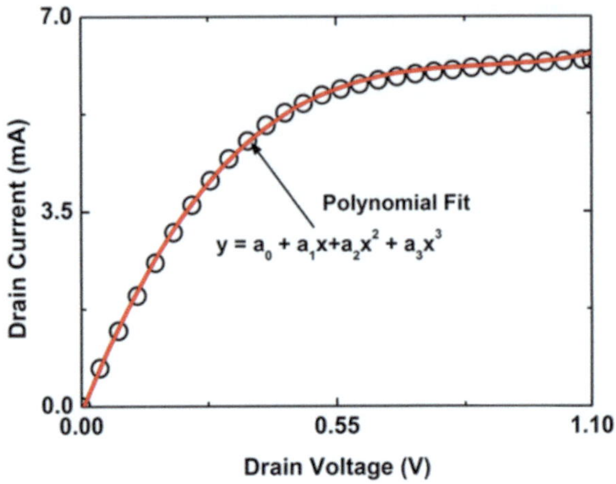

Fig. 2.3 Empirical model uses arbitrary mathematical functions to model the data

Polynomial fit to the measured non-linear data is an example of an empirical model. The polynomial or any other chosen function would typically include an array of coefficients. These coefficients are obtained by fitting measured data. These models can be quite effective. They eliminate the issue of the need of interpolation from the table-based models. However, the issue of the need of extrapolation may not be solved in these models. This is because as the mathematical formulation is developed from data within a boundary of input conditions, its behavior outside the input boundary cannot be assumed to be physical.

2.2.3 TCAD Models

Technology computer-aided design tools allow creation of three-dimensional structure of transistor. These three-dimensional detailed model [24–26] can include a lot of nuanced details about the device. The three-dimensional structure is divided into a large number of mesh points. At each mesh point, fundamental device physics equations are solved. The solutions at all the mesh points are made self-consistent in the simulations. One can immediately see that as a significant detailed information about the device structure is part of the model, these model can capture details about device behavior. An example is shown in Fig. 2.4. The suble changes in structure affect device behavior can be studied with these models. However, as the device needs to be divided into a large number of mesh points and computations performed at each point, one can also see that these models are computationally expensive.

TCAD models are heavily used for device technology development. While there are some mixed-mode approaches where TCAD models are used in circuit simulations too, these are not very popular. The primary reasons are being that TCAD models are too complex for circuit simulations where a solution for a

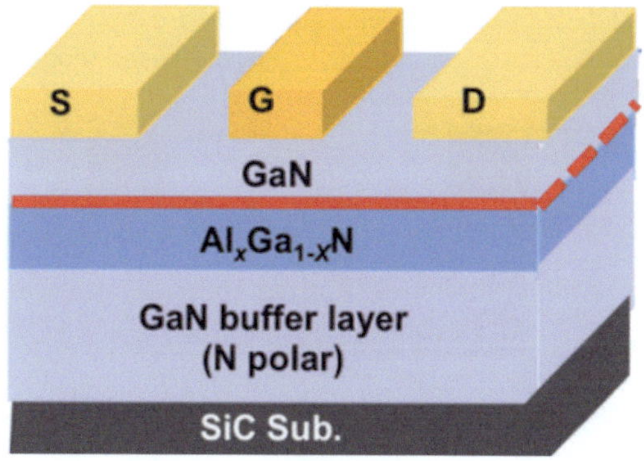

Fig. 2.4 Detailed three-dimensional representation of the device is used in TCAD. This structure is divided into several small parts, and fundamental physical equations are solved at each point

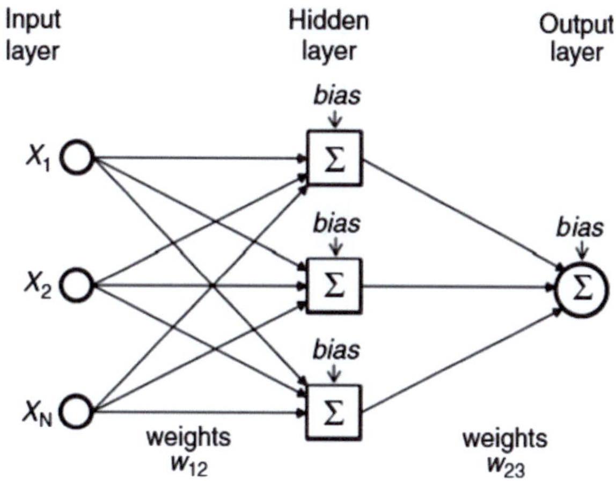

Fig. 2.5 A neural network can be used as a compact model for semiconductor devices

range of input conditions is desired within very short time (typically seconds). So, although TCAD models can be very accurate and are highly predictive of changes in device behavior with its structure, these models are not used for circuit simulations.

2.2.4 *Neural Network Based Compact Models*

Neural networks can be and are used as compact models [27–30]. Neural networks are configurable mathematical models with multiple inputs and outputs. An example of neural network is shown in Fig. 2.5. Compact models of a semiconductor device also have multiple inputs. These include voltage bias, temperature, etc. Compact models also have multiple outputs such as current and capacitances of the device. The relationships between the multiple inputs and outputs can be modeled with neural network. These models do not originate from the physics of the device behavior but use the configurability of the neural networks. The number of hidden

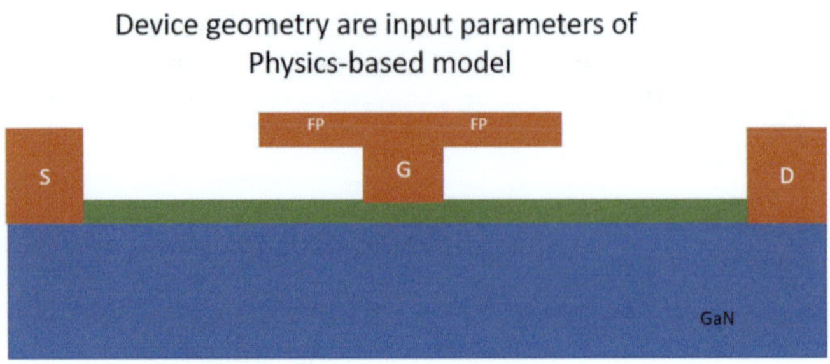

$$I_{ds} = \frac{\mu_0 C_g}{\sqrt{1 + (\theta_{sat} \cdot V_{ds})^2}} \cdot \frac{W}{L} \cdot (V_{go} - \psi_m + V_{th}) \cdot \psi_{ds}$$

Expression derived based on device physics

Fig. 2.6 An example of physics-based compact model showing formulation derived from device physics

layers, neurons in each layer, and weights of individual neurons are configurable in neural network based models.

Neural network formulations also do not originate from physics, similar to empirical models. The main difference and advantage of the neural network over the empirical models is that the underlying mathematical formulations in neural network are continuous and smooth. Neural network suffers from similar disadvantages as the empirical models. For instance, neural networks' scalability w.r.t. geometry and temperature is limited to the measured data used to create the model. If scaled beyond the measurements, neural network cannot guarantee behavior.

2.2.5 Physics-Based Compact Models

Physics-based compact models are developed from fundamental device physics [31–36]. Figure 2.6 is shown as an example of physics-based compact model. These include device electrostatics, carrier distribution, and transport. The device physics equation invariables are interdependent. As a result, analytical formulations for fundamental device equations are difficult. This difficulty is overcome by making justifiable approximation during the development of the model. For instance, it is a quite common approximation to assume a constant carrier mobility during the development of a compact model. Later, this assumption is lifted by introduction of model parameters related to the effect of mobility dependence on electric field.

Physics-based compact models usually have two components: 1. core model and 2. real device effect models. The core formulations are derived from fundamental device physics. These formulations are derived assuming ideal device. For instance, two-dimensional effects in channel are neglected to derive analytical formulations. Once core model has been developed, many real device effects are incorporated in the modeling formulations by appropriate means. Typically, this means the introduction of parameters related to the effects. These parameters are physically meaningful being related to physical effect in the device.

Physics-based compact models are extremely popular in several device technologies. BSIM, PSP, and EKV are examples of physics-based compact models in silicon transistors. ASM and MVSG are examples of physics-based compact models for GaN-based transistors. When developed carefully, physics-based compact models solve the issue related to both interpolation and extrapolation needed during model usage. As the mathematical formulations in these models are derived from device physics, they behave physically in regions model has not been calibrated in. Furthermore, as the parameters introduced in these models are physically meaningful, they are easier to extract from the measurements. Physically meaningful parameters are also easier to check if the extracted value is within reason. These properties are crucial for good performance of the model in circuit simulations.

2.3 Comparison Among Different Compact Models

There are several desirable properties which the compact models should possess. These properties are present in different styles of compact models to varying degrees. These properties often require trade-off. If one tries to achieve more of one, it is easy to lose on the other. These properties are:

- Computational speed: Computational speed of a compact model relates to how fast it simulates on a given computing resource for a given type(s) of simulation. The faster the model the better it is for circuit simulations. Compact models need to be reasonably fast as no designer would like to wait for minutes to hours to simulate the effect of a small change in the design. Design is an iterative process, and many combination of electrical components and their values needs to be simulated. So the speed of simulation is quite important for ease of design tuning. The computational speed of compact model has two aspects to it. First is related to the complexity of model formulations itself. The larger the number of formulations, more complex and computationally inefficient the model is. Another aspect of the speed is related to the type of mathematical functions used in the compact model. Algebraic functions can be computed faster than exponential (e^x), power (x^y), and the trigonometric functions ($tanhx, tanx$ etc.). These computationally expensive functions should be used sparingly in the compact models. Overall, a good computational speed is a key requirement of a compact model.
- Global accuracy: Accuracy is another important requirement from a compact model. Accuracy is measure of how closely a compact model can represent the actual experimental data on the electrical device. For instance, for a transistor,

output and transfer characteristics is a typical measurement to characterize its behavior. Compact model can be simulated in the same manner as the measurements and the simulation results are compared with the measurements to see the accuracy of the compact model. During the model development and parameter extraction, measured data is fitted by tuning the model parameters. The more accurately a compact model can fit the device characteristics the better.

- Predictability: Design process needs tuning and simulating various scenarios for getting the most optimal performance. As a result, the tunable parameters of the compact models are usually tuned and design re-simulated to see the change in performance. A compact model should be able to support these design experiments. This is possible when compact models can predictably simulate the change in electrical behavior with input parameter change. It is easy to see that a physics-based compact model will have better predictive nature as compared to empirical or measurement-based models.
- Robustness: Compact models are used in many types of simulations as needed by circuit design. These simulations can be DC, small-signal AC, transient, or large signal non-linear harmonic balance simulations. Compact model needs to behave well and should not cause non-convergence in all of these simulations. The convergence behavior of a compact model is dependent on its mathematical robustness. For instance, harmonic balance simulations will use several higher order derivatives of current formulations of a transistor model. If the model is mathematically ill-conditioned, for instance has very sharp derivatives, it may lead to non-convergence of harmonic balance simulator.

All of these requirements or qualities of compact models depend on the compact model architecture or type. This is summarized in Fig. 2.7. Figure 2.7 compares three types of compact models for desirable features of speed, robustness, predictability, and accuracy. TCAD models are very accurate and can be highly predictive of the changes made to the device as they capture significant details

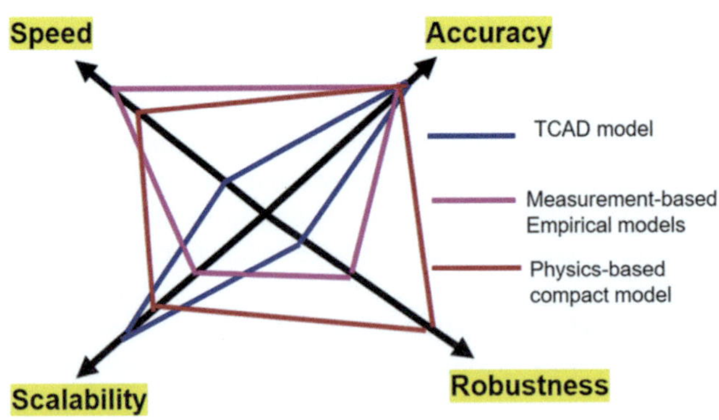

Fig. 2.7 A comparison on how different types of compact models compare for different desirable features

about the device structure. However, their speed is poor and they can also have non-convergence behavior. Non-convergence in TCAD models can occur as the simulator needs to find a self-consistent solution at a large number of mesh points. Measurement-based and empirical compact models can have very good computational speed as they can use simple formulations. However, being created from a set of measurements, they cannot have predictability. Furthermore, the convergence robustness of empirical and measurement-based compact models can be problematic due to several reasons. First, to reach to a solution, simulator can apply inputs that can be well beyond the input ranges used to create these models. As a result, model behavior can be erratic leading to non-convergence. Second, creation of these models relies on quantities that can be measured easily. For instance, terminal capacitances. Now, building a model based only on terminal capacitances may not conserve charge, a physical requirement for good convergence. These issues cause non-convergence in measurement-based and empirical compact models. Physics-based compact models, when carefully created, can strike a right trade-off among these requirements. As they are derived from device physics, they can simulate unforeseen conditions gracefully. Also, fundamental physics is preserved such as charge conservation by design in these models. This leads to good robustness and convergence in these models. If implemented carefully, physics-based compact models can have a good computational speed even though they usually have more complex formulations than empirical models. These models can have a good predictability as they are derived from fundamental device physics. Overall, physics-based compact models cover the largest space of all the key properties as shown in Fig. 2.7. It is for this reason they are quite popular for silicon and now also for GaN technology. Compact Model Coalition (CMC) is a consortium of several leading semiconductor companies around the globe. CMC is an industry standardization body for compact models. Physics-based compact models are preferred by the CMC for the discussed reasons.

2.4 GaN Compact Models

Compact modeling for GaN HEMTs is especially challenging for several reasons. The complexity of GaN device operation and the plethora of non-linear effects in it raise several challenges in GaN compact model development. For instance, the heavily used Boltzmann approximation for the charge formation in the GaN HEMT is not applicable due to the quantum nature of two-dimensional electron gas in HEMTs. The access regions behave highly non-linearly and play an important role in device characteristics. Additionally, there can be multiple field-plates of different configurations in a GaN device. Field-plates also affect device characteristics and need to be modeled. The bias ranges for the input voltages and temperatures are also much larger for GaN device. This means that mathematical equations in the compact models should be carefully formulated and implemented to avoid any numerical problems. These complex issues make the development of a high quality compact model for GaN HEMTs a challenging task.

Nevertheless, the challenges also present opportunities and compact modeling for GaN HEMTs is an area of very intense research. There are many approaches that have been developed by researchers for modeling GaN HEMTs. These different approaches generally fall into the categories of compact models described earlier in the chapter. Compact models for GaN have evolved from measurement-based empirical models to threshold voltage, charge- and surface-potential based physical compact models. This section briefly reviews some of the works on GaN compact models in the literature under the categories of model discussed earlier.

- Measurement-based compact models: Several measurement-based models for small- and non-linear large signal models for GaN HEMTs have been proposed. Root et al. proposed X-parameter based model for GaN HEMTs in [37]. This model uses X-parameters measured at multiple bias points to model the non-linear behavior of the GaN device. A load-line based approach has been proposed by Raffo et al. in [17]. This model uses low-frequency large signal measurement to characterize the resistive core or the current generator of the model. Capacitive effects are modeled on top of the resistive core to model the high frequency behavior of the device.
- Empirical GaN compact models: Several empirical GaN compact models have been developed by the researchers. Angelov model is one of the very popular models. This model has been implemented in several commercial tools. This model uses empirical functions to model the drain current, gate current, and terminal capacitances. The model has also been extended to include trapping effects. Many variants of Angelov model have also been proposed to address needs related to scaling, frequency domain improvements, and other aspects of the model. EEHEMT is another popular GaN compact model that has been implemented in commercial simulators. This model provides formulations to model frequency dependent behavior of the device with additional parameters.
- Physics-based GaN compact models: Physics-based compact models is a new emerging trend for GaN devices. Several researchers have published on such models. Author's works on modeling 2-DEG [38], surface-potential [39], and terminal charges [39] led to the industry standard ASM-HEMT or ASM-GaN compact model [40, 95]. Trew et al. have developed a zone-based model [41]. In this model, the GaN HEMT is divided into different regions. Regional physics is modeled to develop a complete model. Patrick et al. have developed a surface-potential based model in [42]. Ujwal et al. have developed a charge-based GaN HEMT model in [43]. This model resulted in the industry standard MVSG. ASM-HEMT and MVSG are the two industry standard compact models for GaN RF and power transistors. A brief description of model standardization and the Compact model Coalition (CMC) is given in the next section.

2.5 Compact Model Coalition and Industry Standard GaN Compact Models

Compact Model Coalition (CMC) is a consortium of many leading semiconductor companies. These companies include foundries, EDA companies, design houses, and electronics product companies. CMC is the industry standardization body for the compact models for device technologies relevant for their members. To date, there are about 20 standard compact models for various devices and technologies declared as standard by the CMC.

Earliest models which CMC standardized for CMOS technologies. Later, CMC also declared standard models for SOI, FinFET, and FDSOI technologies. Toward the end of 2013, CMC launched a process to select an industry standard model for GaN HEMTs. Several researchers working on GaN compact models were invited, based on their published works. Table 2.1 [44] lists the researchers who were invited by the CMC to participate in industry standardization as a potential candidate.

CMC has a multi-phase rigorous process to select an industry standard compact model. In each phase, CMC filtered the number of candidate models passing to the next phase based on their evaluation criteria. From Table 2.1 four model candidates moved to the next phase, i.e., phase-2. These four were: ASM-HEMT, MVSG, Angelov, and HSP model. After phase-2, two model candidates moved to phase-3. These were ASM-HEMT and MVSG compact models. In the next phase, both ASM-HEMT and MVSG were declared industry standard GaN compact models. In the later chapters of this book, ASM-HEMT model will be described in detail.

Table 2.1 Researchers invited by the CMC as potential candidates for industry standard model [44]. The organizations listed are the affiliations of the authors at the time of the invite as given in [44]

Model name	Invited researcher	Organization
ASM-HEMT/ASM-GaN	Dr. Sourabh Khandelwal	Norwegian University of Science and Technology
MVSG	Dr. Ujwal Radhakrishna	Massachusetts Institute of Technology
HSP	Dr. Patrick Martin	CEA Leti
Angelov	Dr. Itchov Angelov	Chalmers University
HKUST	Dr. M. Chan	HongKong University of Science and Technology
RPI	Dr. M. Shur	Renesselaer Polytechnic Institute
NCSU	Dr. R. Trew	North Carolina State University
UConn	Dr. Anwar	University of Connecticut

Chapter 3
Introduction to ASM-HEMT Compact Model

3.1 User-Level View of ASM-HEMT Compact Model

A compact model for user point of view can be viewed as a black-box with certain inputs and producing the terminal characteristics of the device as the output. This view of the ASM-HEMT model is shown in Fig. 3.1. With ASM-HEMT model as the black-box, the inputs to this black-box include the bias or the input voltages. The bias voltages are the terminal gate, drain, source, and substrate or body voltage applied to a GaN device. Ambient temperature is another input.

The model parameters are divided into four categories:

- Device Geometry Parameters
- Model configuration parameters
- Device physics parameters
- Smoothening model parameters

These are described below in more details.

3.1.1 *Device Geometry Parameters*

Device geometry parameters: These parameters are the dimensions of relevant features in the GaN device. Examples of parameters in this category include: Gate-length L, Gate-width W, Barrier layer thickness T_{bar}, etc. GaN device has many more relevant dimensions which are needed in physics-based compact model. An example device is shown in Fig. 3.2. In Fig. 3.2, the device is divided into five different regions. Region-1: The region under the gate contact is considered as main transistor and the key geometry for this region are L_G and T_{bar}. Region-2 is the region under the gate overhang. This region is considered as the gate field-plated region. Key device geometry parameters are the length of the overhang L_{FP1} and

S. Khandelwal, *Advanced SPICE Model for GaN HEMTs (ASM-HEMT)*,
https://doi.org/10.1007/978-3-030-77730-2_3

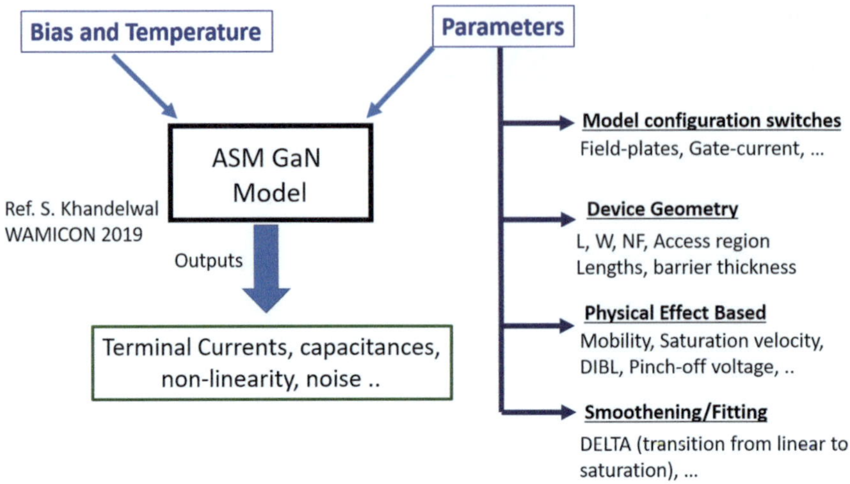

Fig. 3.1 User-level view of the ASM-HEMT compact model. Inputs to the model are bias, temperature, and model parameters. Outputs from the model are terminal current, capacitances, noise, and non-linear behavior of the device

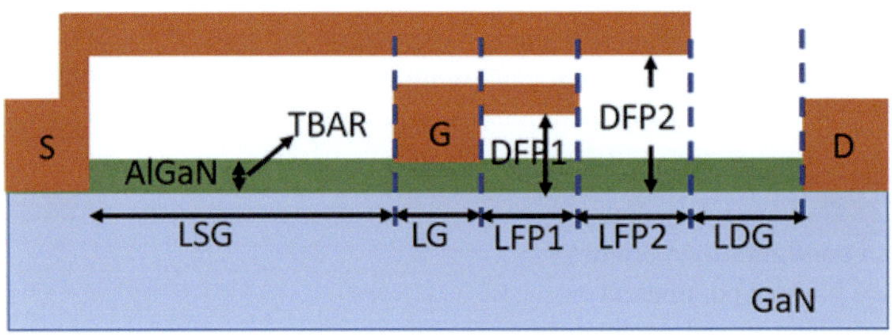

Fig. 3.2 A sample GaN HEMT device cross-section showing the various device dimensions which are model parameters

the distance of the gate overhang from the 2-DEG D_{FP1}. Region-3 is the channel region under the source field-plate or the source field-plated region. The geometry parameters of this region are similar to the ones for the gate field-plate region. Region-4 and Region-5 are the access regions toward the source- and the drain-contact respectively. The length of these access regions is the device geometry parameter.

Figure 3.2 is an example. ASM-HEMT model is re-configurable to model devices with up-to four field-plates on either side of the gate. This ability of the model to re-configure comes from the model configuration switch parameter described below.

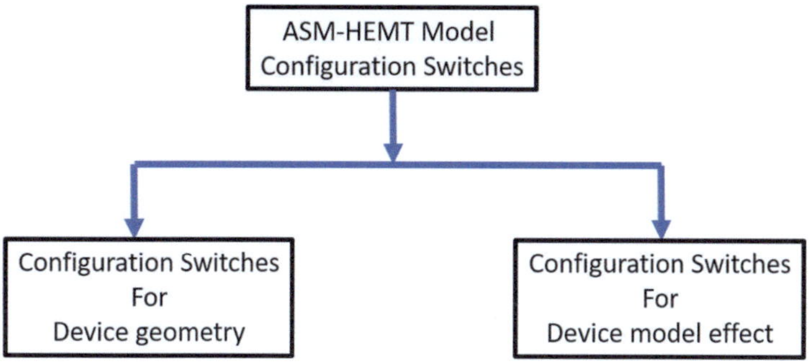

Fig. 3.3 Configurability in ASM-HEMT model is divided into two parts

3.1.2 Model Configuration Switches

These model parameters enable user to configure the model according to their device technology and simulation needs. More detailed impact of these switches on the device behavior is described in a later chapter. Here we describe the idea behind model configuration parameters. GaN HEMTs geometry can vary significantly from device to device. Different end applications need widely different device architecture. For instance, high voltage power GaN device will have more field-plates as compared to a high frequency GaN device. ASM-HEMT model can be configured according to the device using the model configuration switches. In a similar way, the need to model specific effect is also driven by the end application. For instance, flicker and thermal noise models are not crucial for power amplifier application. ASM-HEMT model allows user to turn-off such parts of the model if they are not required through the model configuration switches. The model configuration switches in ASM-HEMT model can be divided into two categories: switches for geometry, and switches for specific model accuracy and effect as shown in Fig. 3.3.

Model Configuration Switches for Device Architecture

For GaN HEMTs the architecture of field-plates i.e. number and configuration of field-plates depends on the device manufacturer and application. ASM-HEMT model can be configured for range of device from zero field-plates to up to total four field-plates on either side of the gate with the help of configuration switches $FPxMOD$ and $FPxSMOD$, where x varies from 0 to 4. Turning $FPxMOD$ on, adds field-plates in the device model. The configuration switches for field-plates are shown in Fig. 3.4.

An example is shown in Fig. 3.5. From source to drain the various regions of the device are marked with dotted lines and labeled as regions. The first region is labeled as R1, and similarly others. In this device R1 is source-connected field-plate, R2

	Model Switch	Description
Drain-side	FP1MOD	Drain-side first field-plate: 0, 1, 2
	FP2MOD	Drain-side second field-plate: 0, 1, 2
	FP3MOD	Drain-side third field-plate: 0, 1, 2
	FP4MOD	Drain-side fourth field-plate: 0, 1, 2
Source-side	FP1SMOD	Source-side first field-plate: 0, 1, 2
	FP2SMOD	Source-side second field-plate: 0, 1, 2
	FP3SMOD	Source-side third field-plate: 0, 1, 2
	FP4SMOD	Source-side forth field-plate: 0, 1, 2

Fig. 3.4 Field-plate model configuration switches are used to configure the model according to the field-plates present in the device being modeled

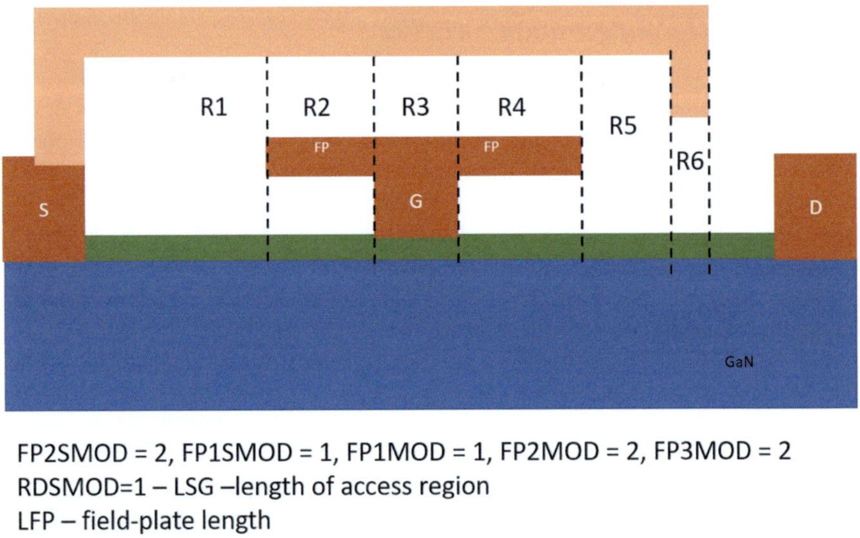

Fig. 3.5 An example field-plate configuration in a GaN HEMT device. This device is modeled with the shown FPMOD combination of switches in ASM-HEMT model

is gate-connected field-plate, R3 is the main channel region, R4 is gate-connected field-plate, and R5 and R6 are source-connected field-plate regions. The field-plate model switches have been set according to this device structure. FP2SMOD = 2 is used for R1, FP1SMOD is used for R2, FP1MOD is used for R4, FP2MOD is used for R5, and FFP3MOD is used for R6. It is to be noted here that as more field-plate regions are included in the model, the complexity of model calculations increases. So it is important to understand the trade-off between increase in model complexity adding additional field-plates in the model and the increase in model complexity. In cases where small field-plate region does not affect overall device behavior, it is better to keep model complexity lower and not introduce a field-plate for that region. For instance R6 in Fig. 3.5.

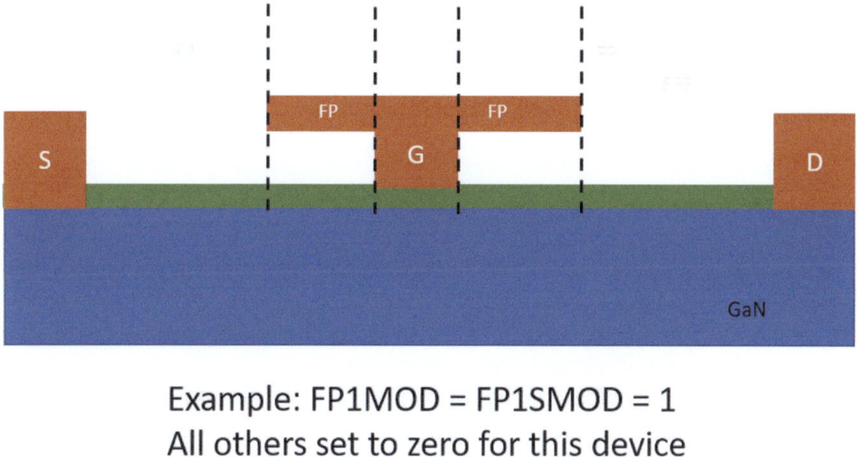

Fig. 3.6 An example of GaN device with different structure as compared to Fig. 3.5. Field-plate configuration for this is also shown

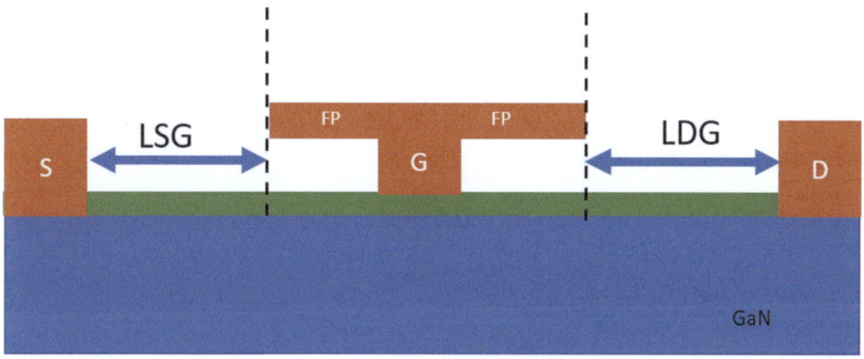

Fig. 3.7 Access regions are typical in GaN HEMTs. A self-aligned manufacturing process gets rid of the access regions. ASM-HEMT can be configured to a self-aligned GaN HEMT device by setting access region length to zero

Another example of the device is shown in Fig. 3.6. This is a different device with only a T-type gate structure. The overhang of T shape can be modeled as field-plate regions in the device. As a result FP1SMOD and FP1MOD are configured in the model of this device as shown in Fig. 3.6. All other field-plate model switches are set to zero for this device.

Another structural difference seen in GaN HEMTs for devices from different manufacturers is for the access regions. Access regions in the device are shown in Fig. 3.7. Although this is less common, however, some GaN devices may not have access regions. This is possible in the device manufacturing with a self-aligned process. A few manufacturers have shown such developments. ASM-HEMT model can be configured to such device configuration. This is possible by setting the access region lengths to zero in the model. Access region length parameter toward source side is LSG and toward the drain side is LDG. Setting these parameters to zero removes the semiconductor part of the access region resistance. The contact resistance at the source and drain is still present in the model.

Model Configuration Switches for Physical Effects

Simulation needs may also drive user selection on the model configuration switches. For instance, even for a device with two field-plates one may decide to not enable any of the $FPMODs$, if one is sure that model will never be used in the regions where the effect of field-plates becomes important.

Another need for model configuration switch arises from the basic trade-off of accuracy versus computational efficiency in a compact model. For instance, $GATEMOD$ is model configuration switch for gate current model in ASM-HEMT model. GATEMOD = 1 is a simple gate current model, while GATEMOD = 2 is a more accurate but complex formulation. If user understands the accuracy needs of his or her application, he or she may choose GATEMOD = 1 as needed and sufficient for his or her application, trading off accuracy for better computational performance.

The model configuration switches are for either turning off a specific physical effect completely, or using a simple formulation or a complex but more accurate formulation for the physical effect. Such model switches are described in the Fig. 3.8.

Model Switch	Description
SHMOD	Self-heating effect 0: Off, and 1: On
GATEMOD	Gate current model 0: Off, 1: Simple model, 2: Advanced model
RGATEMOD	Gate resistance model 0: Off, 1: Simple model, 2: Advanced model
TNMOD	Thermal noise model 0: Off, 1:On
FNMOD	Flicker noise model 0: Off, 1: On
TRAPMOD	Model for trapping effects 0: Off, 1: RF device asymmetric time-constants, 2: RF device symmetric time-constants, and 3: Power device

Fig. 3.8 Switches that affect modeling of physical effects in ASM-HEMT model

3.1.3 Device Physics Parameters

These parameters are related to physical effects in the device. Carrier transport physics related parameters are low-field mobility and saturation velocity. Similarly, physical parameter related to device contacts are contact resistivity. In device modeling practice these parameters are extracted from device measurements. However, the range of these parameters is very well-known. For instance 2-DEG low-field mobility is around 2000 cm^2/V.sec. This is a very important feature of physics-based compact model. Physical parameter has well-known ranges that make extraction easier and also check any errors in measurements.

3.1.4 Smoothing Parameters

Approximations are used in physics-based compact models to capture the complex physics as accurately as possible in a computationally efficient manner. These approximations lead to introduction of model parameters which can be termed as smoothing or fitting. For instance, the transition from linear to saturation region of output characteristics of HEMT is modeled by the parameter $MEXP$ in the ASM-HEMT model. After exercising the model a few items or for few devices, one can get a very good understanding of typical ranges of these parameters.

This section provided a discussion on the user-level view or the black-box view of the ASM-HEMT model. Next we go inside the black-box to see the components in it.

3.2 Components of ASM-HEMT Compact Model

ASM-HEMT model at a high level can be divided into the following key components: 1. Core model, 2. Real device effects model, and 3. Additional model capabilities as shown in Fig. 3.9. These components are discussed below:

- Core model: Core model is the solution of the Schrodinger's and Poisson's equations in the quantum 2-DEG well of the GaN HEMT. Core model also consists of carrier transport physics in the device. Core model formulations assume an ideal device. This means the device is assumed to be having no contact resistances, no local heating, carrier mobility is assumed to be constant with electric field, and device electrostatics is assumed to be one-dimensional. This idealistic assumptions are necessary to derive analytical computationally efficient formulations needed in a useful compact model. While these assumptions are necessary, one can see that real devices do not care about these assumptions!
- Real device effects model: Real devices have a plethora of physical effects which need to be modeled. These models are build or created on top of the

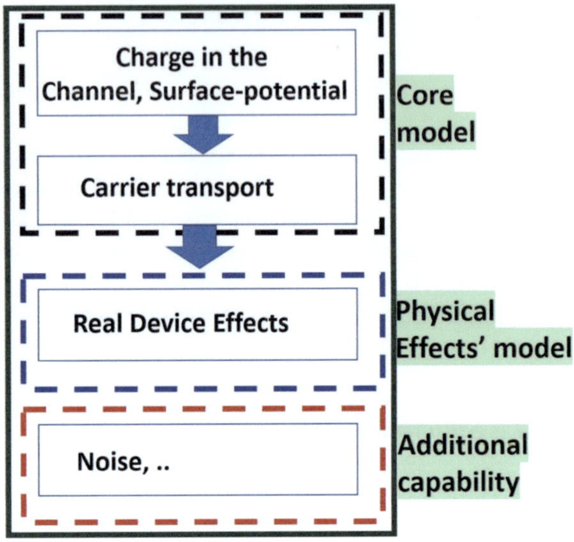

Fig. 3.9 Key components of ASM-HEMT model are core model, real device effects, and additional model capability

core formulations in the ASM-HEMT model. The key idea is to modify the inputs to the core formulations derived for the ideal device to account for specific real device effects. There are many real devices effects and a complete chapter is dedicated in this book to discuss them. Here a simple example is given to illustrate the point on modeling of real device effects. One of the simplifying approximation for derivation of ideal or core device formulation is that the contacts to the terminal of the device have no resistance. As a result, the input voltages to the core formulations are externally applied voltages. To account for the finite contact resistance, instead of calculating core formulations with external voltages, these calculations are performed with intrinsic voltages. Intrinsic voltages V_{dsi} and V_{gsi} account for the drop in contact resistance as shown by formulations below.

$$V_{dsi} = V_{ds} - I_{ds} \cdot (R_s + R_d) \tag{3.1}$$

$$V_{gsi} = V_{gs} - I_{ds} \cdot R_s. \tag{3.2}$$

In Eqs. 3.1, 3.2 V_{ds} and V_{gs} are the externally applied drain-to-source and gate-to-source voltages and I_{ds} is the current in the device. This is a simple explanation of how the core or ideal formulations are updated to account for real device effects. Note GaN HEMT have non-linear access region resistance and these are modeled in the ASM-HEMT model. Here the example is taken to illustrate the idea behind modeling of real device effects in general.

- Additional model capability: Additional model capability in latest version of ASM-HEMT model includes the models for the flicker noise and thermal noise in

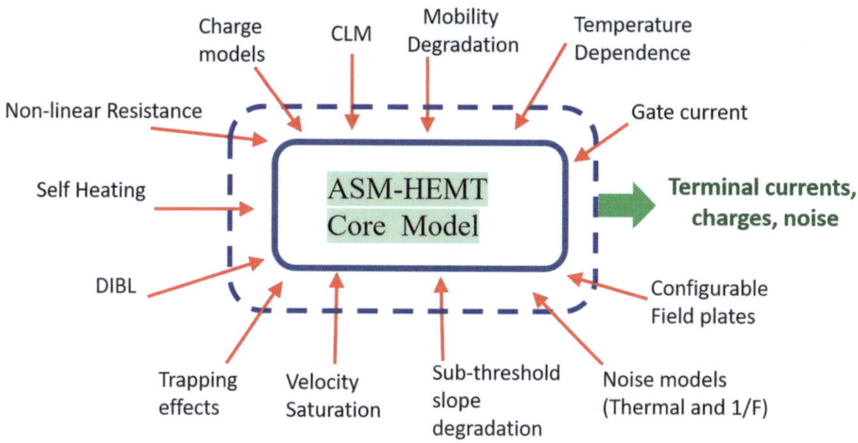

Fig. 3.10 Key components of ASM-HEMT model are core model, real device effects, and additional model capability

the GaN HEMT. These additional capabilities are also modeled self-consistently with other physical aspects such as surface-potential and charges in the device.

Another way to look at the complete model is shown in Fig. 3.10. The core model is the central component of ASM-HEMT model, and a large number of real device effects are modeled with core model as the central platform. This completes the description of key components of the ASM-HEMT model. We discuss model features in the next section.

3.3 ASM-HEMT Model Features

The model features are capability to model and simulate various aspect of GaN device operation. For instance, the ability to model thermal noise of the device is a model feature. Some models may not include this in their implementation. ASM-HEMT has several features making it a full non-linear large signal RF model for GaN RF devices. This model is also featured to model a GaN power device, covering its I-V, C-V, transient, and other relevant aspects of its behavior.

The implemented model features in the ASM-HEMT model can be enabled with the help of model selector switch parameters. A list of model features with the corresponding model selector switch parameter is given in Table 3.1. The possibility of enabling and disabling various model features is a technique used in several compact models.

This way of compact model implementation allows flexibility to users to configure the model in a manner most useful for their application. This is useful for two reasons. First, a few model features may be completely irrelevant to the

Table 3.1 ASM-HEMT model features and corresponding model selector switch to enable the feature

Feature	Model selector	Comments
Non-linear access resistance	RDSMOD	0: Fixed, and 1: Bias-dependent access region resistors
Gate Current	GATEMOD	0: Simple and 1: Advanced gate current model
Thermal noise	TNMOD	Channel and resistance thermal noise
Flicker noise	FNMOD	Channel and gate diode flicker noise
Trapping model	TRAPMOD	Three trapping models
Field-plate drain side	FP1MOD	0: No field-plate, 1: Gate-connected field-plate, and 2: Source-connected field-plate. First field-plate on the drain side
Field-plate drain side	FP2MOD	0: No field-plate, 1: Gate-connected field-plate, and 2: Source-connected field-plate. Second field-plate on the drain side
Field-plate drain side	FP3MOD	0: No field-plate, 1: Gate-connected field-plate, and 2: Source-connected field-plate. Third field-plate on the drain side
Field-plate drain side	FP4MOD	0: No field-plate, 1: Gate-connected field-plate, and 2: Source-connected field-plate. Fourth field-plate on the drain side
Field-plate source side	FP1SMOD	0: No field-plate, 1: Gate-connected field-plate, and 2: Source-connected field-plate. First field-plate on the source side
Field-plate source side	FP2SMOD	0: No field-plate, 1: Gate-connected field-plate, and 2: Source-connected field-plate. Second field-plate on the source side
Field-plate source side	FP3SMOD	0: No field-plate, 1: Gate-connected field-plate, and 2: Source-connected field-plate. Third field-plate on the source side
Field-plate source side	FP4SMOD	0: No field-plate, 1: Gate-connected field-plate, and 2: Source-connected field-plate. Fourth field-plate on the source side
Self-heating	SHMOD	0: No self-heating, 1: Self-heating effect
Gate resistance	RGATEMOD	0: No gate resistance, 1: Gate resistance included

application at hand. For instance, a radio-frequency power amplifier application may not care so much about the flicker noise modeling. So, he can turn the feature completely off. As a result the formulation in the flicker noise model will not be calculated, saving some calculations. Secondly, as for GaN HEMTs the device technology can be quite different from user to user; the model switches allow model configuration for specific technology. For instance, one GaN technology may not use field-plates at all and so the field-plate formulations can be completely turned off. This clearly avoids several unnecessary computations and improves the model performance. Another useful aspect of model configuration switches is the study of impact of various physical effects of electrical performance. For instance, by turning off the non-linear resistance model by RDSMOD switch, one can study how much impact it has on I-V, C-V, or even large signal RF performance of a power amplifier. The switches work as an excellent means for diagnosis.

In summary, model configuration switches provide a very convenient means to tailor the model to the application and the technology it is used for. ASM-HEMT model has several configuration switches as summarized in Table 3.1.

Chapter 4
Core Formulations in ASM-HEMT Model

4.1 Introduction

The main challenge in development of a physics-based formulation of the core of HEMT device operation is the transcendental physics at play. The first goal is to calculate the charge in the quantum well or the two-dimensional electron gas density to develop the model. Two-dimensional electron gas density and the device electrostatics as governed by Poisson equation are interdependent as shown in Fig. 4.1.

4.2 Two-Dimensional Electron Gas Model

Two-dimensional electron gas (2-DEG) formation at the interface of the barrier layer and the GaN layer in GaN HEMT is the core of device operation. This is shown in Fig. 4.2. The source of the 2-DEG in GaN HEMTs was discussed by Ibbetson et al. [45]. They propose the surface stated to be the source of the 2-DEG electrons in GaN HEMTs. The electron in 2-DEG is occupied in a quantum well at the AlGaN/GaN interface. The energy levels of this quantum well are shown in Fig. 4.2.

By applying the gate voltage one can move the Fermi level E_f shown in Fig. 4.2. As a result, the amount of 2-DEG is controlled leading to the transistor action. This gate voltage control of the 2-DEG is the core of the GaN transistor operation. To model the transistor, 2-DEG and its gate voltage control need to be modeled accurately.

The modeling of gate voltage dependence of 2-DEG is challenging because of the interdependent physics which govern this behavior. The quantum well at the AlGaN/GaN interface has multiple energy levels; the 2-DEG is composed of electrons in these energy levels. The number of electrons in each energy levels is dependent on the electric field, and the electric field is dependent on the total

S. Khandelwal, *Advanced SPICE Model for GaN HEMTs (ASM-HEMT)*,
https://doi.org/10.1007/978-3-030-77730-2_4

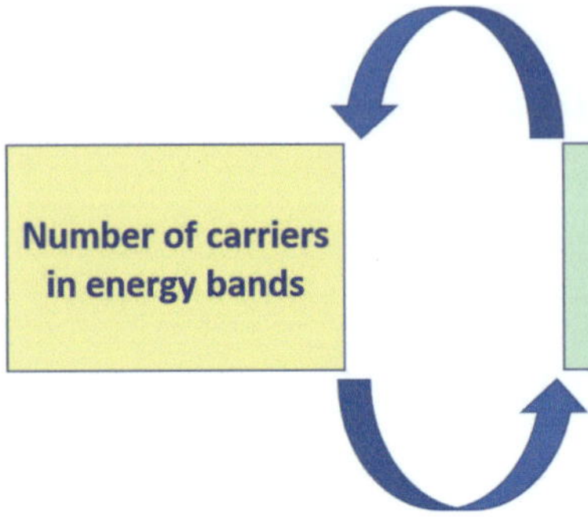

Fig. 4.1 Interdependence of the number of carriers and device electrostatics poses the main challenge in the development of core model

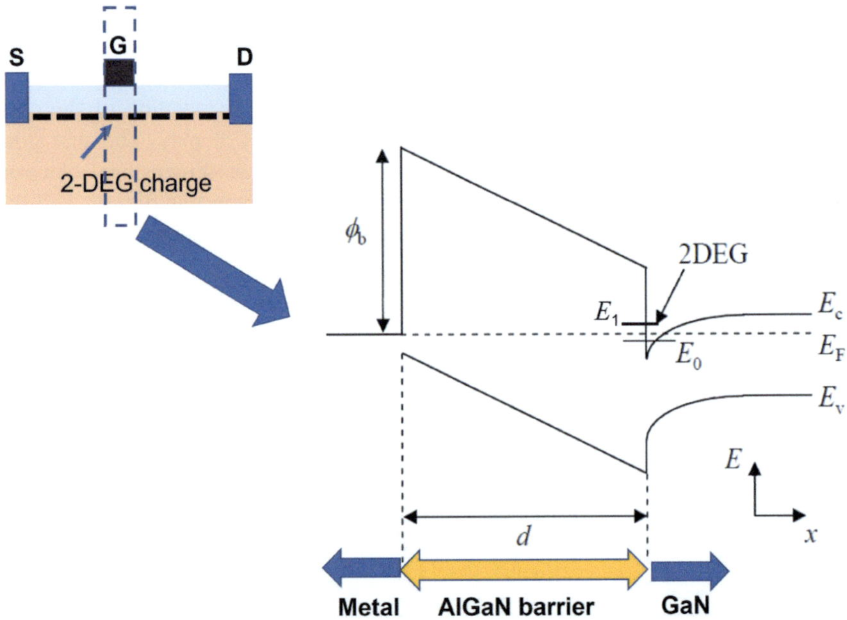

Fig. 4.2 GaN HEMT device and energy band diagram under the gate

number of electrons. This interdependence governed by Poisson's and Schrodinger's formulations needs to be solved self-consistently to develop an accurate model of the 2-DEG density.

The main problem in calculating 2-DEG density in these devices arise is due to the complicated variation of the Fermi level E_f with applied gate and channel potential. A self-consistent solution of the Schrodinger's and Poisson's equations in the quantum well, assuming triangular profile for the potential is expressed as [46],

Table 4.1 Description of symbols used

Symbol	Physical meaning/expression
n_s	Density of 2-DEG
E_f	Position of Fermi level
E_0, E_1	Position of first- and second-energy level
d	Thickness of AlGaN layer
D	Density of states
q	Electron charge
V_{th}	Thermal voltage
ϵ	Permittivity of AlGaN
C_g	ϵ/d Gate Capacitance per unit area
$\gamma_{0,1}$	Experimentally determined parameters
	2.12e-12, 3.73e-12 V.m$^{4/3}$ [21]

$$n_s = DV_{th}\left[\ln\left(\exp\frac{E_f - E_0}{V_{th}} + 1\right) + \ln\left(\exp\frac{E_f - E_1}{V_{th}} + 1\right)\right] \tag{4.1}$$

$$E_{0,1} = \gamma_{0,1} n_s^{2/3} \tag{4.2}$$

$$n_s = \frac{\epsilon}{qd}\left(V_{go} - E_f - V_x\right), \tag{4.3}$$

where $V_{go} = V_g - V_{off}$ and V_x is the channel potential at any point x in the channel. A description of all the other symbols used is given in Table 4.1. It is apparent that to obtain an analytical solution from Eqs. 4.1–4.2, we need to make simplifications. This is done by subdividing the variation of E_f versus the applied gate voltage into different operating regions, allowing us to derive explicit expression in each one. This subdivision is determined by the position of the Fermi level E_f relative to the energy levels E_0 and E_1. The various regions of the subdivision are identified in Fig. 4.3, together with the numerical solution for E_f, E_0 and E_1 versus V_g. The regions are: (i) $V_g < V_{off}$, the sub-V_{off} region, where $|E_f|$ is comparable to $|V_{go}|$; (ii) $V_g > V_{off}$, and $E_f < E_0$, the moderate 2-DEG region; (iii) $V_g > V_{off}$ and $E_f > E_0$, the strong 2-DEG region.

In the sub-Voff region, $|E_f| \gg E_0$ and E_1 (see Fig. 4.3) and therefore $E_f - E_0$ and $E_f - E_1 \approx E_f$. Applying this observation and using the approximation $ln(1+x) \approx x$ for $x << 1$ to Eq. 4.1, and invoking 4.2, the solution for ns can be written in terms of a Lambert function as,

$$\frac{q \cdot n_s}{C_g \cdot V_{th}} \cdot exp\left(\frac{q \cdot n_s}{C_g \cdot V_{th}}\right) = \frac{2qD}{C_g} exp\left(\frac{V_{go}}{V_{th}}\right). \tag{4.4}$$

In the sub-Voff region where $qns << C_g V_{th}$, we obtain the following explicit function for n_s:

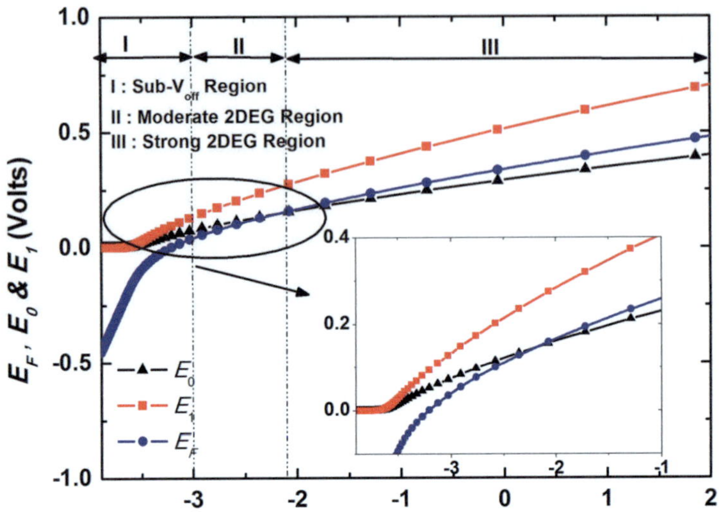

Fig. 4.3 The change in E_f with gate voltage. The regions for compact model development are determined based on E_f crossing E_0 and E_1

$$n_{s,sub-V_{off}} = 2D \cdot V_{th} \cdot exp\left(\frac{V_{go}}{V_{th}}\right). \tag{4.5}$$

This expression is valid when $|Ef| >> E_0$, E_1 and $E_f < 0$.

The model for n_s valid in the moderate 2-DEG and strong 2-DEG regions can be written as,

$$n_{s,above-V_{off}} = \frac{C_g \cdot V_{go}}{q} H(V_{go}), \tag{4.6}$$

where

$$H(V_{go}) = \frac{V_{go} + V_{th}[1 - ln(\beta V_{gon})] - \frac{\gamma_0}{3} \cdot \left(\frac{C_g \cdot V_{go}}{q}\right)^{\frac{2}{3}}}{V_{go} \cdot \left(1 + \frac{V_{th}}{V_{god}}\right) + \frac{2\gamma_0}{3} \cdot \left(\frac{C_g \cdot V_{go}}{q}\right)^{\frac{2}{3}}}. \tag{4.7}$$

Here, V_{gon} and V_{god} are functions of V_{go} given by the interpolation expression,

$$V_{gox} = \frac{V_{go}\alpha_x}{\sqrt{V_{go}^2 + \alpha_x^2}}, \tag{4.8}$$

where, $\alpha_n = e/\beta$ and $alpha_d = 1/\beta$.

The equations from Eq. 4.5 to 4.6 are a solution for 2-DEG charge density valid in specific regions. However, for the purpose of developing a compact model, a continuous unified expression for n_s valid in all regions of operation is desirable.

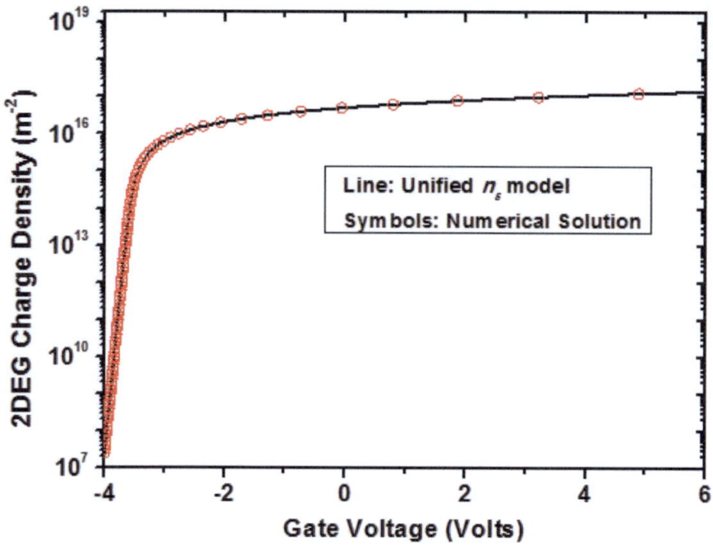

Fig. 4.4 Unified model of n_s of Eq. 4.9 models the variation of 2-DEG charge density in all regions of device operation

The unified analytical expression should also be mathematically continuous for the higher order derivatives of the charge. Such an expression is developed by combining Eq. 4.5 to 4.6 in the following manner:

$$n_{s,uni} = \frac{2V_{th}C_g/q \cdot ln(1 + exp(V_{go}/V_{th}))}{1/H(V_{go,p}) + (C_g/qD)exp(-V_{go}/2V_{th})}. \tag{4.9}$$

This analytical formulation models the variation in 2-DEG charge density accurately for all the regions. A comparison of this model with numerical simulations is shown in Fig. 4.4.

The developed formulation for n_s serves as a starting point for surface-potential calculation as described in the next section.

4.3 Surface-Potential Model

The surface-potential calculation needs very accurate calculation of the Fermi level E_f. From Eqs. 4.1–4.3 and the approximation explained in the n_s model, one can obtain the model for E_f is sub-voff region to be,

$$E_{f,sub-V_{off}} = V_{go} - \frac{2qDV_{th}}{C_g} \exp\left(\frac{V_{go}}{V_{th}}\right). \tag{4.10}$$

Using the function $H(\cdot)$ derived in previous section, E_f for above V_{off} region is written as,

$$E_{f,above} = V_{go}\left(1 - H(V_{go})\right). \tag{4.11}$$

These regional expressions for E_f can be combined to develop a unified solution for E_f as shown below [47, 48],

$$E_{f,unified} = V_{go} - \frac{2V_{th}\ln\left(1 + \exp(V_{go}/2V_{th})\right)}{1/H(V_{go,p}) + (C_g/qD)\exp(-V_{go}/2V_{th})}. \tag{4.12}$$

Here $V_{go,p}$ is equal to V_{go} above V_{off} and is of the order of thermal voltage when $V_g < V_{off}$. It can be observed that when $V_g > V_{off}$ we recover Eq. 4.11 from Eq. 4.12 while in the sub-V_{off} regions $1/H(V_{go,p})$ becomes very small and Eq. 4.10 is obtained. Thus Eq. 4.12 is valid in all the regions, but the accuracy obtained with $E_{f,unified}$ is of the order of milli-volts in the region close to V_{off}. This is because in this region the energy levels E_0 and E_1 (see Fig. 4.3) are very close to each other and both should be considered to calculate E_f. For a surface-potential-based compact model, an accuracy of the order of nano-volts is desirable. So, we refine the solution given by Eq. 4.12 including the contribution from E_1 and obtain an accuracy of the order of femto-volts or better in all the regions. We first compute the quantities $k_{0/1}$, $\xi_{0/1}$, V_{gef}, p, q, and r defined in Table 4.2, and then evaluate E_f given as,

$$E_{f1} = E_{f,unified} - \frac{p}{q}\left(1 + \frac{pr}{2q^2}\right). \tag{4.13}$$

We re-evaluate $k_{0/1}$, $\xi_{0/1}$, V_{gef}, p, q and r with E_{f1} in place of $E_{f,unified}$ and obtain the final solution $E_{f,s}$ from (9). This computation steps follow from Householder's method for solving implicit functions.

The developed E_f model is then used to calculate ψ, which is defined as $E_f + V_x$ [12], where V_x takes the value of the source potential V_s at the source side and the drain potential V_d at the drain side of the channel. The source-side and the drain-

Table 4.2 Expression for quantities in Eq. 4.13

Quantity	Expression
$k_{0,1}$	$\gamma_{0,1}\left(\frac{C_g}{q}\right)^{2/3}$
V_{gef}	$V_g - V_{off} - E_f$
$\xi_{0,1}$	$\exp\left(\frac{E_{f,unified} - k_{0,1}V_{gef}^{2/3}}{V_{th}}\right)$
p	$\frac{C_g}{q}V_{gef} - \sum_{i=0}^{1} DV_{th}\ln(\xi_i + 1)$
q	$-\frac{C_g}{q} - \sum_{i=0}^{1}\frac{D}{1+\xi_i^{-1}}\left(1 + (2/3)k_i V_{gef}^{-1/3}\right)$
r	$\sum_{i=0}^{1}\frac{\frac{2}{9}V_{gef}^{-4/3}Dk_i(1+\xi_i^{-1}) + \frac{D}{V_{th}}\left(1+\frac{2}{3}k_i V_{gef}^{-1/3}\right)^2}{\left(1+\xi_i^{-1}\right)^2}$

side surface-potentials (ψ_s and ψ_d, respectively) calculated using the above method are then used to derive the intrinsic charges as described in the next section.

4.4 Terminal Charge Model

The core charge model calculates the charges at each of the intrinsic device terminals. These terminals are shown in Fig. 4.5.

The intrinsic gate, drain, and source charges in the device have to be calculated accurately for correct transient and capacitance simulations. An accurate modeling of intrinsic charges requires a proper assignment of the channel charge to the terminals of the device. Normally this is done by assigning drain charge as

$$Q_d = x_p Q_{ch} \tag{4.14}$$

and source charge as

$$Q_s = (1 - x_p)Q_{ch}, \tag{4.15}$$

where $Q_{ch} = qn_s = C_g(V_{go} - \psi)$ is the channel charge and x_p is the partitioning factor. To simplify the model, partitioning schemes like 60/40, 50/50, etc. have been proposed in standard compact models. A more rigorous and physical way of dividing Q_{ch} as proposed by Ward [49] is by defining

$$Q_d = \int_0^L \frac{x}{L} Q_{ch}(V_g, V_x).dx \tag{4.16}$$

and

$$Q_s = \int_0^L \left(1 - \frac{x}{L}\right) Q_{ch}(V_g, V_x).dx. \tag{4.17}$$

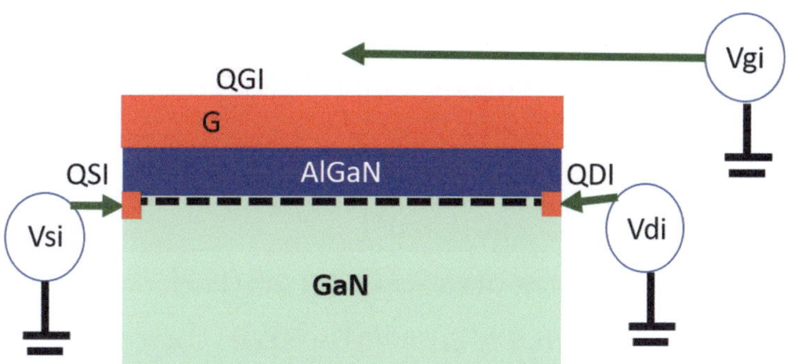

Fig. 4.5 Intrinsic terminals of the GaN HEMT device. Terminal charges for each of the intrinsic terminals are derived in the core model

The total gate charge is then given by

$$Q_g = -Q_s - Q_d. \tag{4.18}$$

Note that by defining the charges in this way, charge conservation is always assured. We use these definitions to derive explicit expressions for all the intrinsic charges.

From the above definitions the total Q_g can be written as,

$$Q_g = -\int_0^L qWn_s(V_g, V_x)dx = -\int_0^L qWC_g\left(V_{go} - \psi(x)\right)dx. \tag{4.19}$$

To perform the integration in Eq. 4.19, we use current continuity and neglect the effect of velocity saturation on charge. We use the surface-potential based drain current model and write,

$$\begin{aligned} dx &= \frac{C_g W\mu(V_{go} - \psi + V_{th})}{I_{ds}} d\psi \\ &= \frac{L(V_{go} - \psi + V_{th})}{(V_{go} - \psi_m + V_{th})(\psi_d - \psi_s)} d\psi, \end{aligned} \tag{4.20}$$

where $\psi_m = (\psi_d + \psi_s)/2$ is the mid-point surface-potential. Applying Eq. 4.20 in 4.19 and performing the integration from source to drain we obtain,

$$\begin{aligned} Q_g = \frac{C_g LW}{\left(V_{go} - \psi_m + V_{th}\right)} \Big[& V_{go}^2 + (1/3)\left(\psi_d^2 + \psi_s^2 + \psi_d\psi_s\right) \\ & -V_{go}(\psi_d + \psi_s - V_{th}) - V_{th}\psi_m \Big]. \end{aligned} \tag{4.21}$$

To calculate the total drain charge 4.20 is integrated from source to an arbitrary point x in the channel resulting in,

$$x = \frac{L(\psi - \psi_s)}{(V_{go} - \psi_m + V_{th})}\left(V_{go} + V_{th} - \frac{\psi + \psi_s}{2}\right). \tag{4.22}$$

Using Eq. 4.22 in the definition of Q_d and performing the integration from source to drain we obtain Eq. 4.23.

$$\begin{aligned} Q_d = -\frac{C_{gLW}}{120(V_{go} - \psi_m + V_{th})^2} \Big[& 12\psi_d^3 + 8\psi_s^3 \\ & +\psi_s^2\left(16\psi_d - 5(V_{th} + 8V_{go})\right) + 2\psi_s\left(12\psi_d^2\right. \\ & \left. -5\psi_d(5V_{th} + 8V_{go}) + 10(V_{th} + V_{go}) \cdot (V_{th} + 4V_{go})\right) \end{aligned}$$

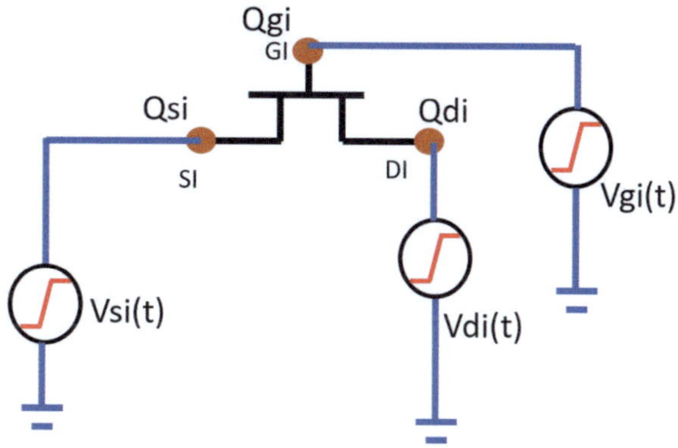

Fig. 4.6 Terminal charges are functions of applied voltages. As any of the voltage changes in time, a terminal charge will change causing a current flow

$$+15\psi_d^2(3V_{th} + 4V_{go}) - 60V_{go}(V_{th} + V_{go})^2$$

$$+20\psi_d(V_{th} + V_{go})(2V_{th} + 5V_{go})\Big]. \quad (4.23)$$

After obtaining Q_g and Q_d, Q_s can simply be calculated using $Q_s = -Q_g - Q_d$ from the charge conservation. This completes the modeling of all the intrinsic device charges. All the nine non-reciprocal intrinsic device capacitances can be obtained from the present charge model by using the definition $C_{ij} = k_{ij}\frac{dQ_i}{dV_j}$, where $k_{ij} = 1$ for $i = j$ and $k_{ij} = -1$ otherwise, with i and j corresponding to the device terminals.

This concept of obtaining capacitance behavior of the device with the charges is explained with the help of Fig. 4.6

The change in any of the applied voltage in time-domain leads to change in the charge associated with all the terminals. This change in charge with time is capacitive current, represented with the mathematical formulations shown in Fig. 4.7. These formulations can be rewritten as shown in Fig. 4.8. It can be seen from Fig. 4.8 that the change in charges with applied voltages leads to capacitances. Furthermore, the capacitances between each pair of terminal arise from the formulations. For instance, from the derivative of gate charge w.r.t source and drain we get C_{gs}, and C_{gd}, respectively. The readers can also appreciate that with this approach the multi-bias dependence of C_{gs} and C_{gd} is included in the model with only one expression of the gate charge. This is a significant advantage over empirical approaches of modeling the capacitance behavior of devices.

$$i_g = \frac{dQ_g}{dt}, i_d = \frac{dQ_d}{dt}, i_s = \frac{dQ_s}{dt}$$

$$i_d + i_s \neq 0$$

$$i_g = \frac{dQ_g}{dV_g} \cdot \frac{dV_g}{dt} + \frac{dQ_g}{dV_s} \cdot \frac{dV_s}{dt} + \frac{dQ_g}{dV_d} \cdot \frac{dV_d}{dt}$$

$$i_s = \frac{dQ_s}{dV_g} \cdot \frac{dV_g}{dt} + \frac{dQ_s}{dV_s} \cdot \frac{dV_s}{dt} + \frac{dQ_s}{dV_d} \cdot \frac{dV_d}{dt}$$

$$i_d = \frac{dQ_d}{dV_g} \cdot \frac{dV_g}{dt} + \frac{dQ_d}{dV_s} \cdot \frac{dV_s}{dt} + \frac{dQ_d}{dV_d} \cdot \frac{dV_d}{dt}$$

Fig. 4.7 Capacitive current flow due to the change in terminal charges if the applied voltage changes with time

$$i_g = \frac{dQ_g}{dt}, i_d = \frac{dQ_d}{dt}, i_s = \frac{dQ_s}{dt}$$

$$i_d + i_s \neq 0$$

$$i_g = C_{gg} \cdot \frac{dV_g}{dt} + C_{gs} \cdot \frac{dV_s}{dt} + C_{gd} \cdot \frac{dV_d}{dt}$$

$$i_s = C_{sg} \cdot \frac{dV_g}{dt} + C_{ss} \cdot \frac{dV_s}{dt} + C_{sd} \cdot \frac{dV_d}{dt}$$

$$i_d = C_{dg} \cdot \frac{dV_g}{dt} + C_{ds} \cdot \frac{dV_s}{dt} + C_{dd} \cdot \frac{dV_d}{dt}$$

Fig. 4.8 Charge-based formulation leads to the definition of terminal capacitances of the device

4.5 Core Drain Current Model

Transistor action can be analyzed by considering two aspects. First is charge control in the channel, and second is the transport of these charges across the channel causing the observed current. Core charge and surface-potential model formulations discussed in the previous section models the charge control process. In order to calculate the current we need to formulate the process of transport of carriers in form of mathematical model. To this end first we need to consider the mechanisms through which carriers are transported in a channel.

4.5.1 Mechanisms of Carrier Transport

There are two primary mechanisms for carrier transport in GaN HEMTs which are included in the ASM-HEMT model. These are drift and diffusion mechanisms of carrier transport. Both these are included in a single model formulation which calculates the drain current in the model.

Drift is the movement of carriers in presence of electric field. When a non-zero drain–source voltage is applied across the device the carriers in the two-dimensional electron gas density experience force. This is shown in Fig. 4.9. The direction of the electric field in the channel is from drain to source as shown in Fig. 4.9. The electrons in the channel flow in the opposite direction from source to drain. This movement of carriers is one component of the current flow.

The carrier movement in presence of electric field is modeled with the concept of carrier mobility. As the applied electric field across the channel increases, the carriers experience more force and hence can move at a higher velocity toward the drain. This change in carrier velocity with applied electric field is modeled with mobility of the channel. The carrier velocity, electric field, and mobility are related as,

$$v = \mu \cdot E. \tag{4.24}$$

Carrier motion within the semiconductor channel is affected by the scattering event they experience as they travel through the channel. The scattering can be due to the other lattice atoms, impurities in the lattice, defects, etc. Nevertheless, the mobility parameter μ captures this phenomenon, and it is a property of the material of

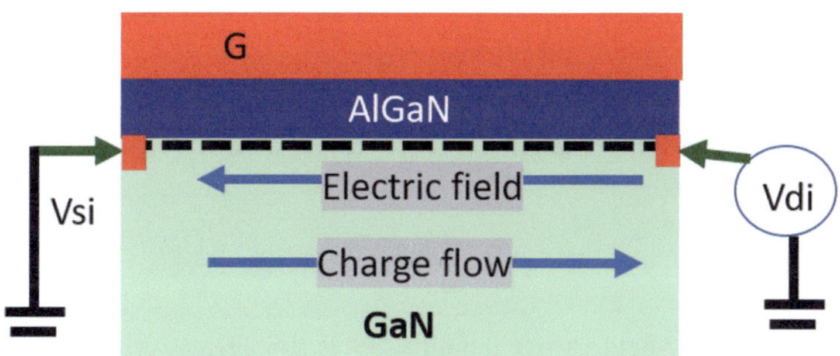

Fig. 4.9 Non-zero drain-source voltage causes electric field in the channel, leading to transport of carriers

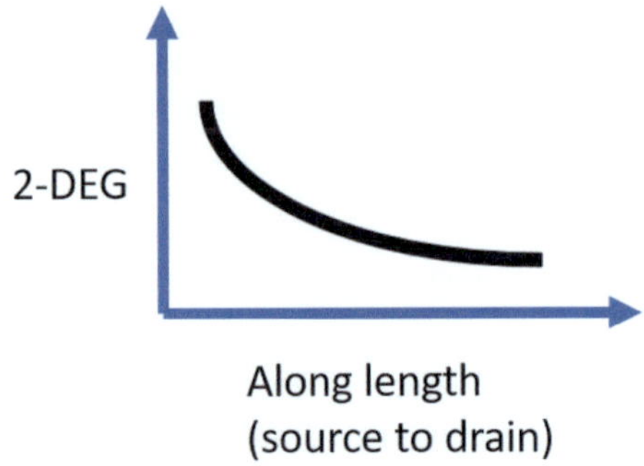

Fig. 4.10 A gradient of carriers in the channel causes diffusion

the channel. Diffusion is the second mechanism with which the carriers transport. Diffusion is due to the gradient in the carrier density. The charge variation with the position in the channel is shown in Fig. 4.10. The carrier density is more at the source than at the drain, and this gradient leads to carrier movement toward the drain, and hence current flow. The current flow due to the diffusion is governed by the charge gradient, and is given by,

$$J(x) \propto \frac{dQ}{dx} \tag{4.25}$$

$$J(x) = D \cdot \frac{dQ}{dx}, \tag{4.26}$$

where D is the diffusion constant of the channel material. As the carriers diffuse, they get scattered for the same reasons as mentioned earlier while describing the drift mechanism. In case of diffusion, one difference is that the thermal energy of the carriers also becomes important. Carrier with large thermal energy can diffuse easier. Overall the diffusion constant D is related to the thermal conditions and mobility by the Einstein relation. The diffusion current density is then given by,

$$J(x) = \mu \cdot V_{th} \cdot \frac{dQ}{dx}, \tag{4.27}$$

where V_{th} is the thermal voltage.

The current at any location x can be written as,

$$I_{ds} = \mu_0 W \cdot Q_{ch} \cdot \frac{d\psi}{dx} + \mu_0 W \cdot V_{th} \cdot \frac{dQ_{ch}}{dx}, \tag{4.28}$$

where μ_0 is the low-field carrier mobility, W is the channel-width, and $Q_{ch} = qn_s = C_g \cdot (V_{go} - \psi)$ is the channel charge at location x. The first term on the right hand side is the drift component, and the second term is the diffusion component of the current. Equation 4.28 can be rearranged as,

$$I_{ds} = -\mu_0 W \cdot \tilde{Q_{ch}} \cdot \frac{d\psi}{dx}, \tag{4.29}$$

where $\tilde{Q_{ch}}$ is given by,

$$\tilde{Q_c}h = V_{th} \frac{dQ_{ch}}{d\psi} - Q_{ch}. \tag{4.30}$$

With increase in the lateral electric field the carriers gain sufficient energy to be scattered by optical phonons, resulting in a decrease in their mobility and ultimately saturation of the carrier velocity. To account for this effect Eq. 4.29 is modified as,

$$I_{ds} = \frac{\mu_{eff} \tilde{Q_{ch}} \frac{d\psi}{dx}}{\sqrt{1 + (\mu_e ff / \upsilon_{sat} E_x)^2}}, \tag{4.31}$$

where υ_{sat} is the saturation velocity of the carriers, E_x is the lateral electric field. Equation 4.31 is integrated from source to drain assuming an average electric field $E_x = (\psi_d \check{} \psi_s)/L$ with L being the channel length of the device. Equation 4.31 can be integrated from source $x = 0$ to the drain $x = L$ of the device. The integration yields,

$$I_{ds} = \frac{\mu_0 C_g}{\sqrt{1 + (\theta_{sat} \cdot V_{ds})^2}} \cdot \frac{W}{L} \cdot (V_{go} - \psi_m + V_{th}) \cdot \psi_{ds}. \tag{4.32}$$

Equation 4.32 is the complete core drain–source current formulation. This model assumes ideal device. For instance, the model derivation did not account for thermal effects, access region resistance in GaN HEMTs, etc. These and several other real device effects need to be accounted for in the model. The modeling of real device effects is discussed in the next chapter.

Chapter 5
Non-ideal Effects in Device Current and Their Modeling

5.1 Introduction

The drain current formulation shown in Eq. 4.32 is formulation for an ideal device. Indeed, several assumptions were made during the derivation of Eq. 4.32. In fact, if these assumptions were not made, it would have been extremely difficult if not impossible to derive an analytical formulation for the drain–source current. These assumptions will be lifted in this chapter and the core drain–source model will be augmented with models for a range of physical effects.

5.2 Pinch-Off Effect

The charge profile in the 2-DEG is tapered as shown in Fig. 5.1. At high drain–source voltage there can be very small charge density at the drain end. This small charge density at the drain essentially isolates the drain from the channel. As a result, effective drain voltage which influences the channel can be different than the applied drain–source voltage. In calculating surface-potential at the drain side this effect needs to be accounted for. This is achieved in ASM-HEMT model as shown in Fig. 5.2.

The surface-potential calculations at the drain side are performed with an effective drain voltage. Effective drain is calculated as shown in Fig. 5.3.

S. Khandelwal, *Advanced SPICE Model for GaN HEMTs (ASM-HEMT)*,
https://doi.org/10.1007/978-3-030-77730-2_5

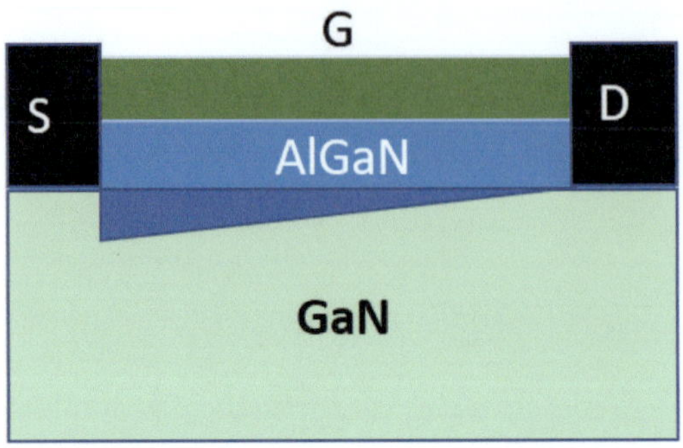

Fig. 5.1 At high drain–source voltage the charge density is tapered and drain voltage is isolated from the channel

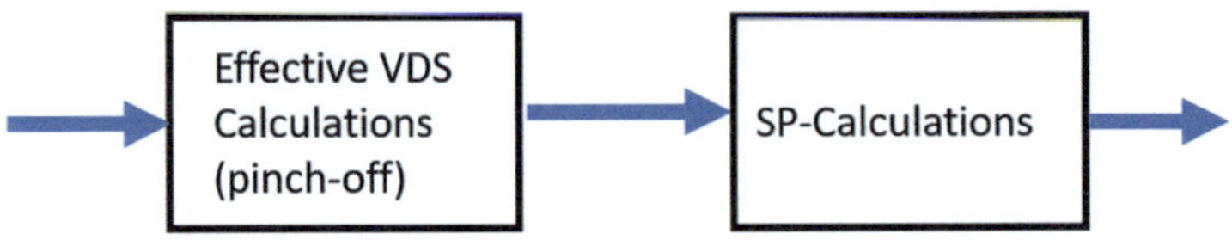

Fig. 5.2 Effective (not applied) drain voltage is used in the core model to account for the pinch-off effect

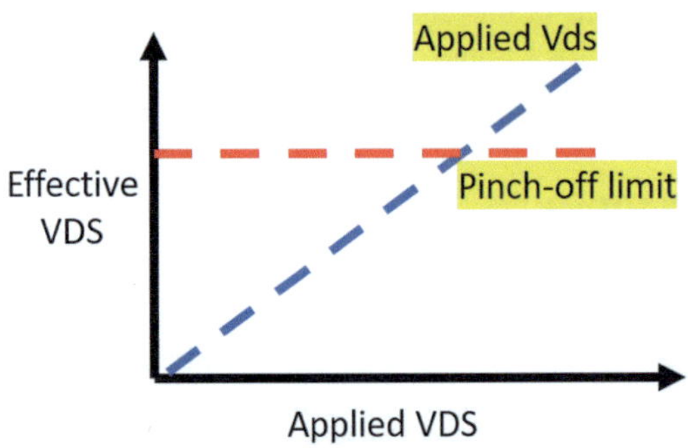

Fig. 5.3 Effective (not applied) drain voltage is used in the core model to account for the pinch-off effect

5.3 Velocity Saturation Effect

As the carrier travel in the channel they scatter due to the presence of other lattice atoms, impurities, defects, etc. With increase in applied drain–source voltage the lateral electric field increases, and carriers move faster. However, the carrier velocity cannot keep increasing with the applied drain–source voltage. As they witness scattering the carrier velocity saturates. The saturation of carrier velocity also means that a more increase in the drain–source voltage is not going to cause any change in the current. To account for this effect, the drain-side surface-potential calculation

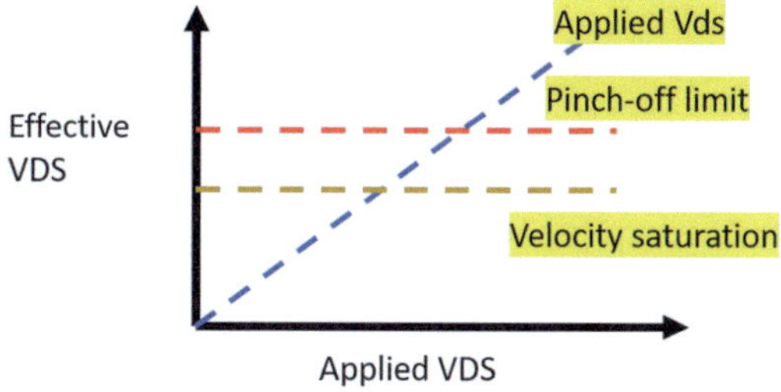

Fig. 5.4 Drain voltage calculation needs to account for the velocity saturation effect

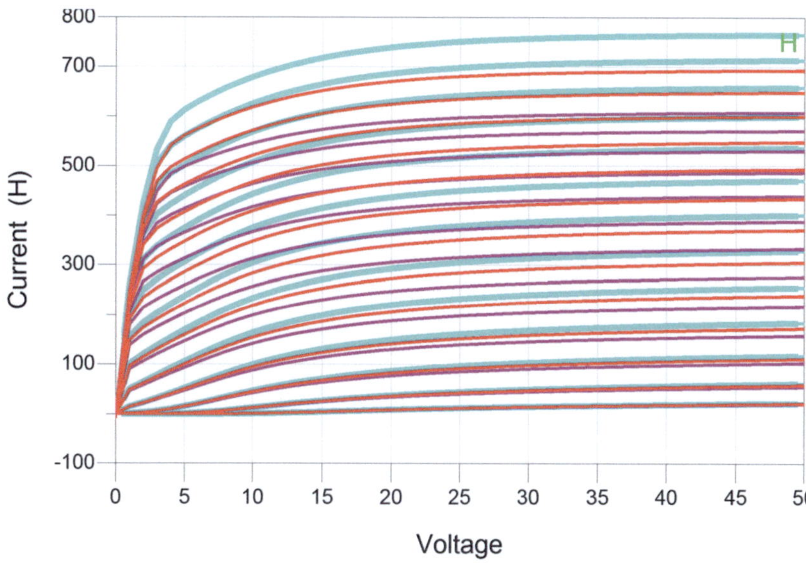

Fig. 5.5 Effect of saturation velocity on I-V characteristics

needs to account for the velocity saturation effect in similar approach as that of the pinch-off voltage. Indeed, this is followed in ASM-HEMT model as shown in Fig. 5.4.The drain–source saturation voltage due to the velocity saturation effect is given by,

$$V_{d,sat} = \frac{vsat \cdot L}{\mu_0}, \tag{5.1}$$

where $vsat$ is the saturation velocity of the carriers, and μ_0 is the low-field carrier mobility. The impact of saturation velocity on the I-V characteristics of the device can be seen from Fig. 5.5.

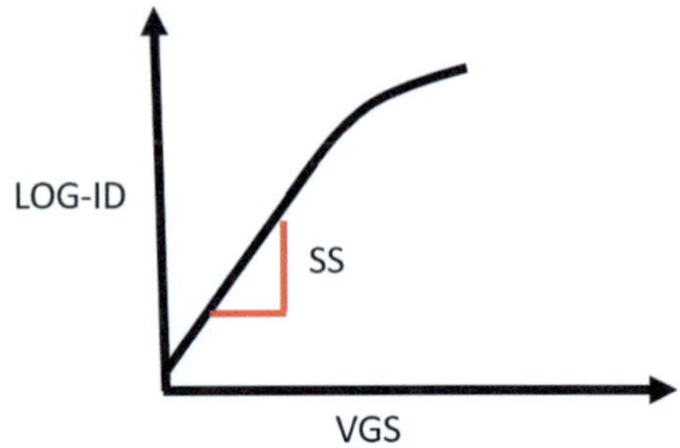

Fig. 5.6 Sub-threshold slope of transistors

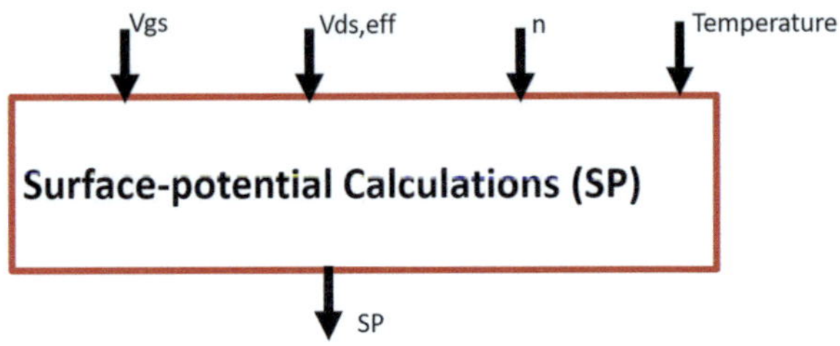

Fig. 5.7 Sub-threshold slope degradation factor is used in surface-potential calculations to account for the effect of non-ideal sub-threshold slope

5.4 Non-ideal Sub-threshold Slope

The slope of logarithm of drain current w.r.t gate–source voltage is known as sub-threshold slope. This is shown in Fig. 5.6. In an ideal device the sub-threshold slope of the device is ideal 60 mV/decade at room temperature. In reality this theoretical value is not found. To account for the non-ideal sub-threshold slope, a sub-threshold slope degradation factor is calculated. This is given by,

$$n = 1 + \mathrm{NFACTOR} + \mathrm{CDSCD} \cdot V_{ds}, \tag{5.2}$$

where NFACTOR and CDSCD are model parameters. The sub-threshold slope degradation factor NFACTOR then is used in the surface-potential calculations at both source and drain to account for this effect as shown in Fig. 5.7. The effect of NFACTOR is shown in Fig. 5.8. The effect of sub-threshold slope parameter CDSCD on the drain current is shown in Fig. 5.9. It can be seen from Figs. 5.8 and 5.9 that NFACTOR affects in all drain–source voltage conditions whereas CDSCD affects I-V for high drain–source voltage conditions.

5.5 Mobility Dependence of Vertical Field

The mobility of the carriers in the 2-DEG channel is assumed to be independent of the gate electrode electric field in the core drain–source current model. However, as the applied gate voltage and hence the electric field increase, the mobility of the

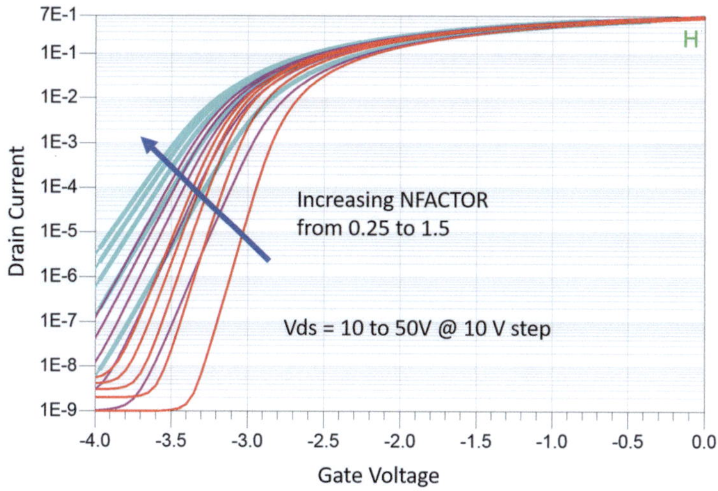

Fig. 5.8 Effect of NFACTOR on I-V characteristics

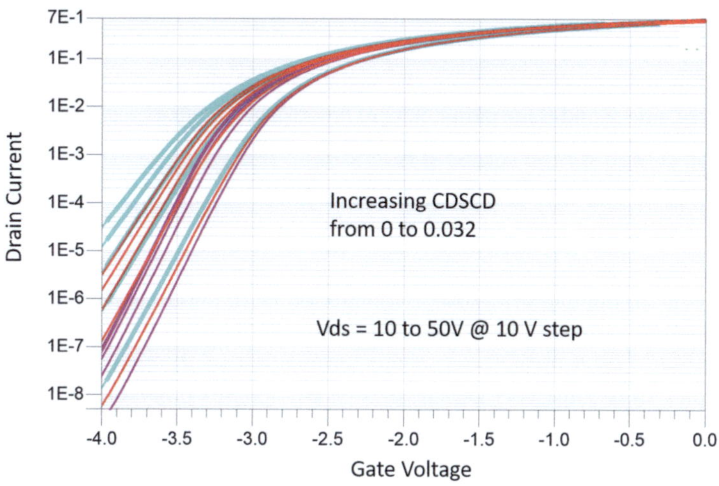

Fig. 5.9 Impact of the sub-threshold slope parameter CDSCD on the device current

carriers decreases. This is shown in Fig. 5.10. This is attributed to increased carrier scattering under the influence of vertical electric field.

There are multiple scattering mechanisms known for silicon transistors. These include Coulomb scattering, phonon scattering, and surface roughness scattering. Coulomb scattering is due to the dopant charges in the channel. GaN HEMTs have undoped channel, and Coulomb scattering is not significant for this reason.

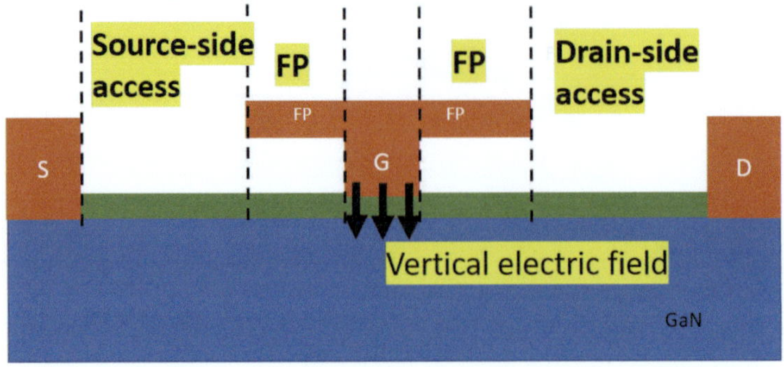

Fig. 5.10 Large vertical electric field can reduce the carrier mobility in the channel

Accounting for the phonon and surface roughness scattering we evaluate effective mobility as,

$$\mu_{eff} = \frac{\mu_0}{1 + \alpha E_x + \beta E_x^2}, \tag{5.3}$$

where μ_0 is the low-field mobility, E_x is the vertical electric field at location x in the channel, and α and β are parameters to be extracted from experimental data. The vertical electric field at location x is given by,

$$\mathrm{e}_x = \frac{V_{gs} - \psi_x}{T_{bar}}, \tag{5.4}$$

where V_{gs} is the applied gate–source voltage, ψ_x is the surface-potential at location x in the channel, and T_{bar} is the barrier layer thickness.

If Eq. 5.4 is used in the core model formulation of the previous chapter, the integration will become very complex. To overcome this problem, average vertical electric field in the channel is used in the effective mobility expression of ASM-HEMT model. The average electric field is given by,

$$\mathrm{e}_x = \frac{V_{gs} - \psi_m}{T_{bar}}, \tag{5.5}$$

where ψ_m is the average surface-potential obtained by averaging the surface-potential at the source and the drain ends of the channel. The effective mobility is then given by,

$$\mu_{eff} = \mu_{eff} = \frac{\mu_0}{1 + UA \cdot E_{avg} + UB \cdot E_{avg}^2}, \tag{5.6}$$

where UA and UB are ASM-HEMT model parameters. The typical values for these parameters are UA $\approx 10^-9$ and UA $\approx 10^-18$. The impact of mobility dependence

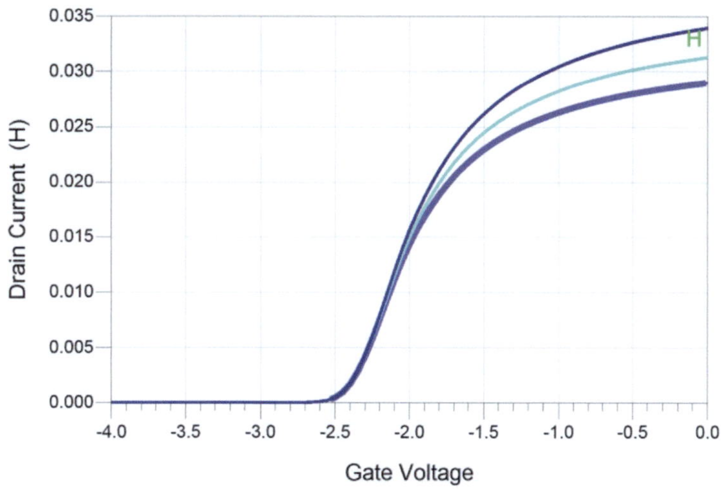

Fig. 5.11 Impact of mobility dependence of vertical electric field on the I-V behavior of GaN-HEMT. This effect is modeled in ASM-HEMT model with UA

on vertical electric field on the I-V characteristics of GaN-HEMT is shown in Fig. 5.11. It can be seen from Fig. 5.11 that at low V_{gs} the effect is not significant; however, for high V_{gs} regions this effect becomes quite significant. The impact of mobility-field dependence as modeled with the help of UA parameters is shown in Fig. 5.11.

5.6 Channel-Length Modulation Effect

A high drain–source voltage leads to a tapered charge profile as mentioned earlier. If drain–source voltage increases further, the taper can get sharper or in other works the location toward the drain where the charge is minimum starts to move toward source. This is shown in Fig. 5.12 with device pictures for three drain–source voltages. Applied drain–source voltage increases from left to right in the picture with VDS1 > VDS2 > VDS3. As seen in Fig. 5.12 the charge profile gets increasingly more tapered, with location of the point of lowest charge moving toward source. This shift in the point of lowest charge with drain–source voltage is termed as channel-length modulation effect. The channel-length modulation effect causes a finite output resistance in the I-V characteristics of the device. The effect on channel-length modulation on the I-V characteristics can be seen from Fig. 5.13.

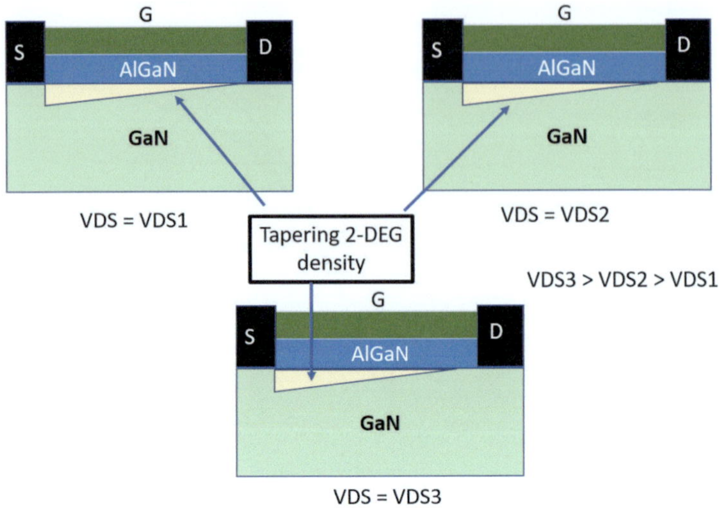

Fig. 5.12 Impact of large drain–source voltage and channel-length modulation effect

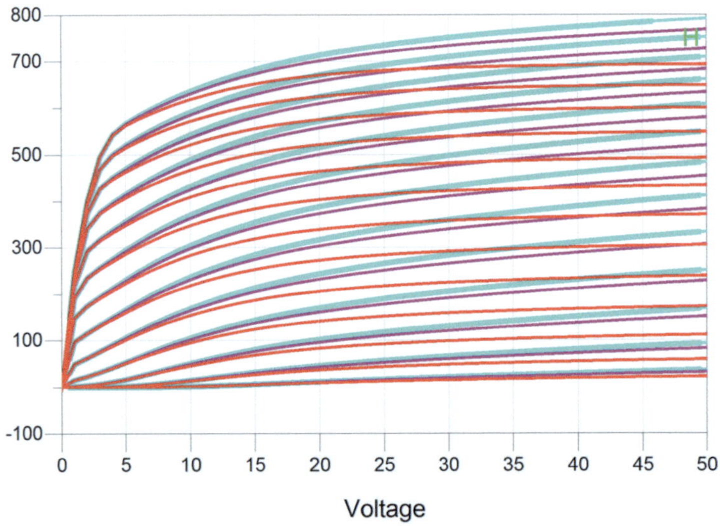

Fig. 5.13 The slope of the output characteristics is affected by the channel-length modulation parameter

5.7 Drain-Induced Barrier Lowering Effect

For large drain–source values, the barrier from electron entry into the channel can be modulated by the drain voltage as well. This is shown in Fig. 5.14. This leads to change in the cut-off voltage with applied drain–source voltage. The impact of

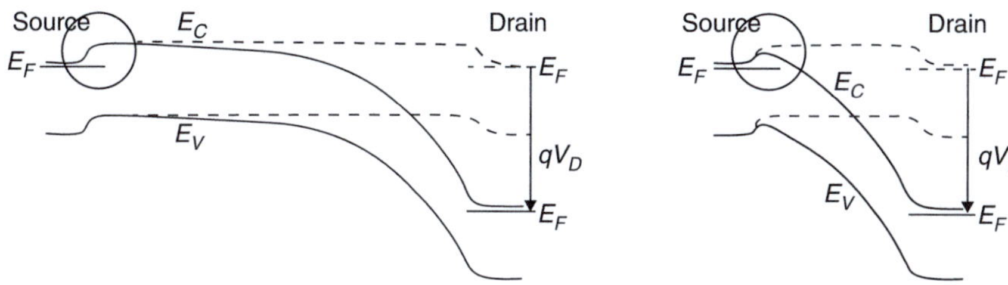

Fig. 5.14 Large drain–source voltage can modify the barrier at the source end of the channel and affect the cut-off voltage

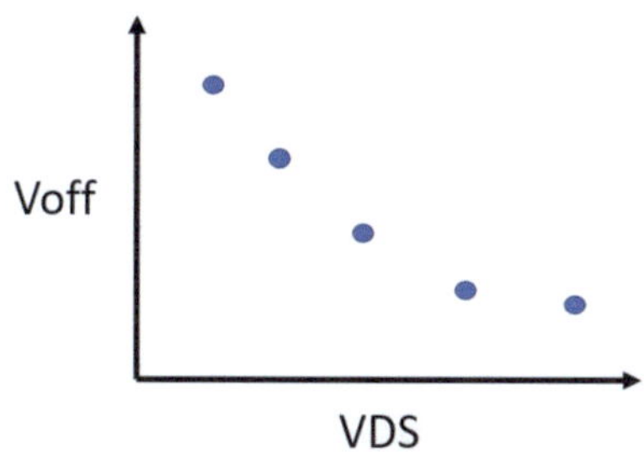

Fig. 5.15 Non-linear variation of cut-off voltage with the applied drain–source voltage is modeled with the help of VDSCALE model parameter

drain–source voltage on the cut-off voltage is found to be non-linear. For instance the observed cut-off voltage for different applied drain–source voltages is found to follow the behavior shown in Fig. 5.15. The non-linearity is modeled with the following formulation.

$$\Delta V_{off,DIBL} = ETA0 \cdot \frac{V_{ds}\mathrm{VDSCALE}}{\sqrt{(V_{ds}^2 + \mathrm{VDSCALE}^2)}}. \tag{5.7}$$

The impact of DIBL effect parameter on the I-V characteristic can be observed from Figs. 5.16, 5.17, and 5.18.

5.8 Non-linear Access Region Resistance Model

Typical GaN HEMTs do not have self-aligned source and drain contacts, and there are access regions at both sides of the gate. This is shown in Fig. 5.19. Access regions behave as non-linear resistors and affect the overall terminal characteristics of the devices. The non-linearity of access regions is modeled in ASM-HEMT model with physical formulations.

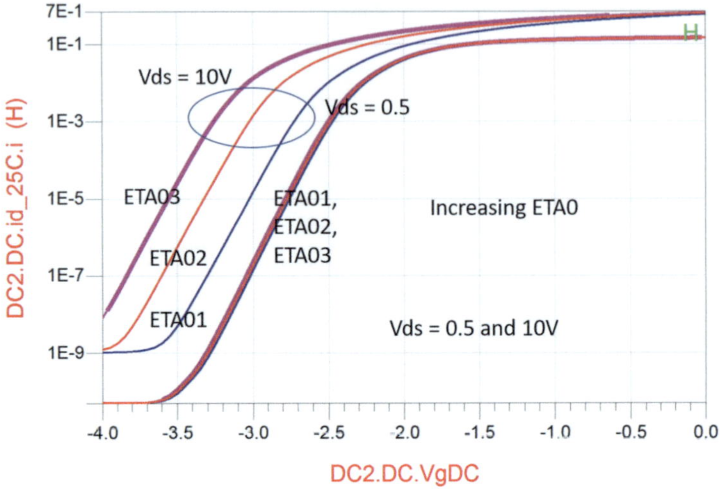

Fig. 5.16 Impact of DIBL effect parameter ETA0 on the transfer characteristics

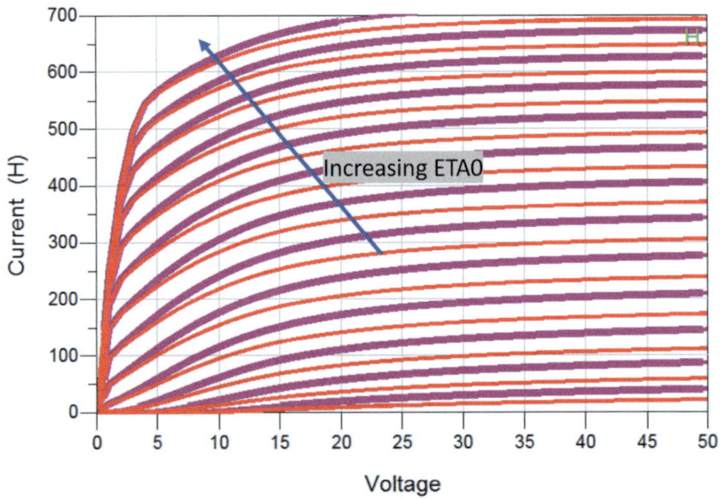

Fig. 5.17 Impact of DIBL effect parameter ETA0 on the output characteristics

The access region has fixed two-dimensional electron density charge, i.e., the 2-DEG charge in these regions is not modulated by the application of gate voltage. When the voltage drop in the access regions is small these regions behave like linear resistors. The resistance value is given by,

$$R_{acc} = \frac{L_{acc}}{qn_{s,acc} \cdot W \cdot \mu}, \tag{5.8}$$

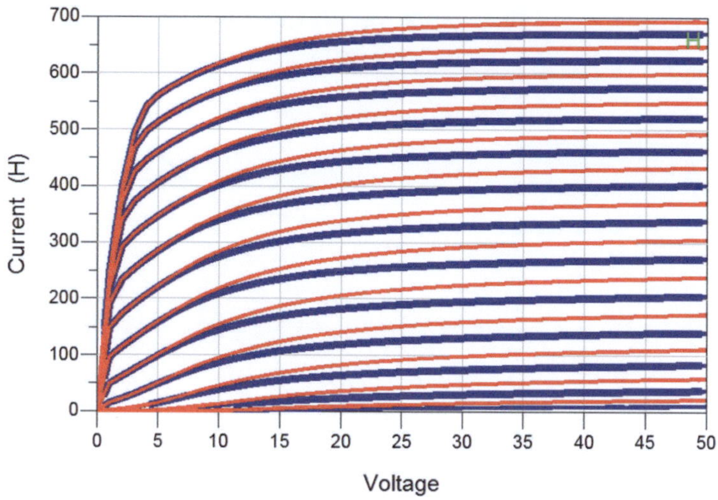

Fig. 5.18 Impact of DIBL effect parameter VDSCALE on the output characteristics

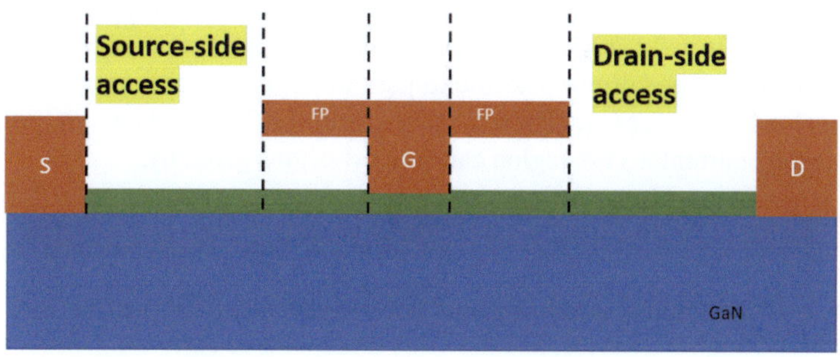

Fig. 5.19 Typical GaN HEMT with access regions toward the drain and source highlighted in the picture

where L_{acc} is the length of the access region, μ is the carrier mobility in access regions, and $n_{s,acc}$ is the 2-DEG charge density in the access region. The carrier moves in the access region with velocity related to the electric field and mobility of carriers as given by,

$$v_{acc} = \mu \cdot E_{acc}, \tag{5.9}$$

where v_{acc} is the velocity of carriers in the access region, E_{acc} is the lateral electric field in this region. E_{acc} can be approximated as

$$E_{acc} = \frac{V_{acc}}{L_{acc}}. \tag{5.10}$$

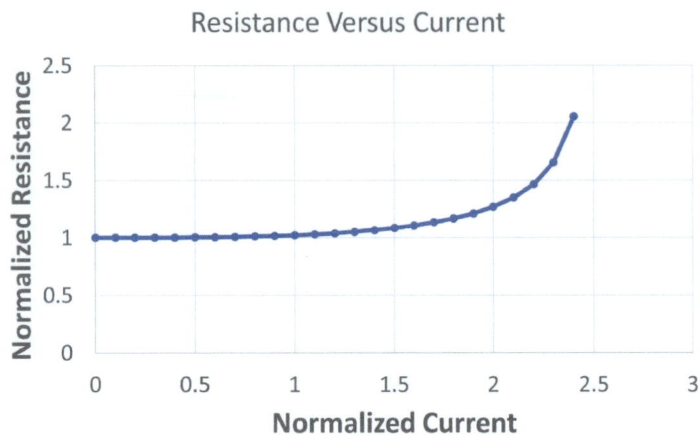

Fig. 5.20 Resistance of the access region increases non-linearly as larger current is made to pass through the region

For small E_{acc} the carrier velocity in the access regions can increase linearly with E_{acc}. However, the velocity saturation phenomenon can occur in access regions too. In such condition, the current possible to pass through the access region saturates, and the value of the access region resistance increases. This is shown in Fig. 5.20.

The largest current access region can support is thus given by,

$$I_{sat,acc} = qn_{s,acc} \cdot v_{sat}, \tag{5.11}$$

where v_{sat} is the saturation velocity in the access regions. The current is of course continuous through the device, and the same drain–source current I_{ds} flows through the access region as through the region under the gate. Considering this, the resistance of the access region is formulated in the following manner in ASM-HEMT model:

$$R_{s,acc} = \frac{LSG}{W \cdot NF \cdot q \cdot \mathrm{NS0ACCS(T)} \cdot \mathrm{U0ACCS(T)}} \times \left(1 - \left(\frac{I_{ds}}{I_{sat,acc}}\right)^{\mathrm{MEXPACCS}}\right)^{\frac{-1}{\mathrm{MEXPACCS}}} \tag{5.12}$$

The access regions also have metal contacts. These metallic contacts have finite resistivity. As a result there is also a contact resistance in the device. The contact resistance is given by,

$$R_{sc} = \frac{RSC}{W \cdot NF}, \tag{5.13}$$

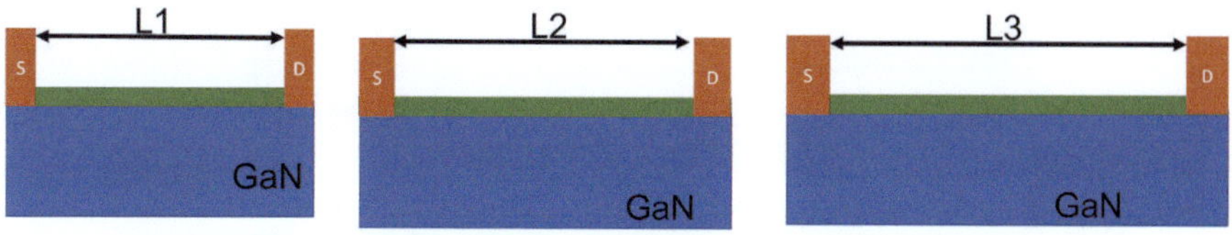

Fig. 5.21 Test structures which can be used for extracting the contact resistance

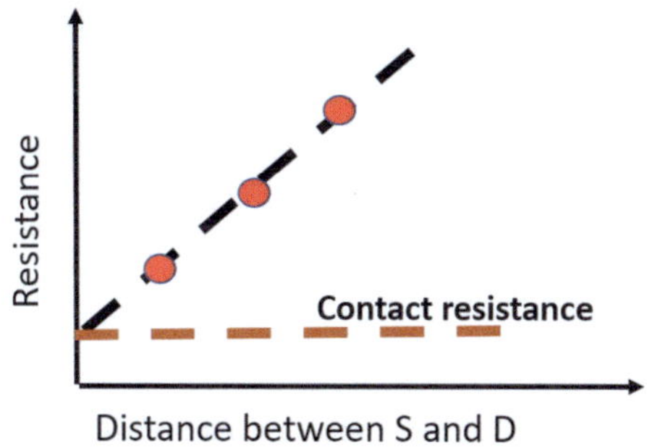

Fig. 5.22 Contact resistance can be extracted by extrapolating the measured resistance of test structures shown in Fig. 5.21

where the model parameter RSC represents the resistivity per square. The total resistance is then given by,

$$R_{s,acc,total} = R_{sc} + R_{s,acc}. \tag{5.14}$$

The contact resistance can be extracted using special modeling test structures. Such structures are shown in Fig. 5.21. In Fig. 5.21 gate has not been placed on the semiconductor. Also structures with varying distance between source and drain have been used.

The contact resistance does not scale with the length of the access region. This can be seen from Eq. 5.13. On the other hand the access region resistance scale with the length, see Eq. 5.12. This scaling difference between the contact and access region resistance is used in extracting the two from the test structures shown in Fig. 5.21. By measuring these test structures for their I-V characteristics their resistance is obtained. Resistance at relatively small applied voltage gives the total resistance shown in Eq. 5.14. The total resistance should increase linearly with the access region length as shown in Fig. 5.22. By extrapolating the total access region resistance to zero access region length, contact resistance can be extracted.

For certain GaN HEMTs there can be an influence of gate electric field on the charge in the access regions. This is shown in Fig. 5.23. This effect is usually very small and is used for only fine tuning of the model. The effect is modeled with K_{0ACC} parameter. The model formulation accounting for change in the access region charge density is,

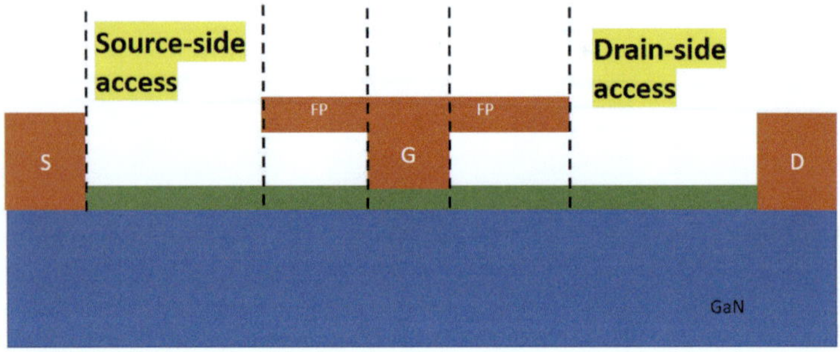

Fig. 5.23 The fringe fields from the overhand of the T-gate can affect the charge density of the access regions

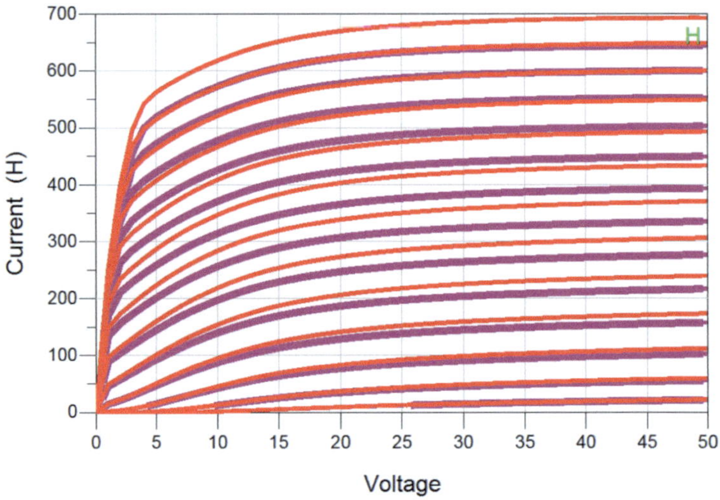

Fig. 5.24 Impact of changing both source- and drain-side contact resistance on I-V

$$n_{acc} = n_{acc0} \cdot (1 + k_{0acc} \cdot V_{gs}). \tag{5.15}$$

The impact of non-linear resistance on I-V model can be seen from Figs. 5.24, 5.25, and 5.26. The contact resistance affects both linear and the saturation part of the I-V characteristics. The affect of contact resistance on linear part of the I-V is obvious. The total resistance of the device between source and drain is given by,

$$R_t = R_{ch} + R_{acc} + R_c, \tag{5.16}$$

where R_{ch} is the channel resistance, R_c is the contact resistance, and R_{acc} is the access region resistance. In linear region R_c directly adds to the R_{ch}. The effect of R_c in the saturation region is somewhat more involved. R_{ch} in the saturation region is high, and R_c is relatively low saturation region. However, it still affects

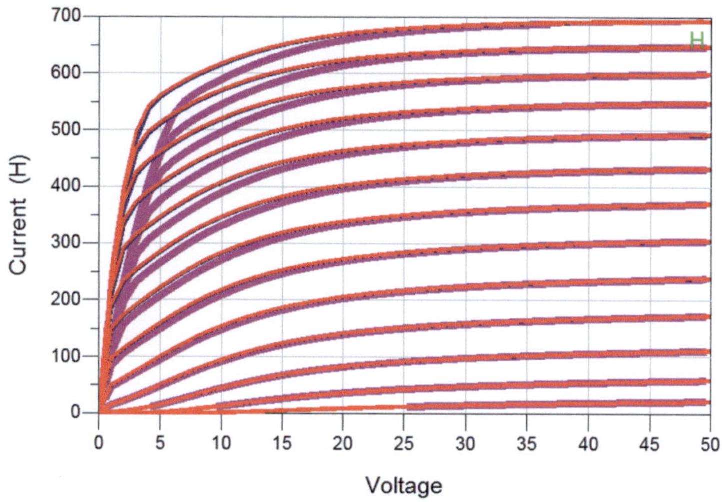

Fig. 5.25 Impact of changing drain-side contact resistance only on I-V

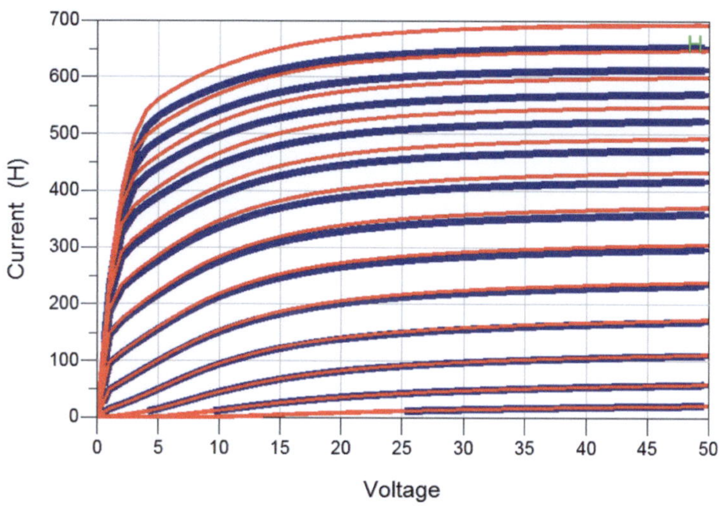

Fig. 5.26 Impact of changing access region charge density I-V

the saturation region because of the source-side contact resistance. The current flowing through the source-side resistor causes the gate–source voltage to reduce. This reduces the saturation region resistance even after relatively low values of the R_c. This can be clearly seen when Figs. 5.24 and 5.25 is compared. In Fig. 5.25 only the drain-side contact resistance is varied, and it can be seen that it impacts only in the linear I-V region.

The effect of the 2-DEG charge density in the access region on I-V behavior is shown in Fig. 5.26. The impact of access region 2-DEG charge density is different than the contact resistance. The effect is governed by the fact that the access region resistance is small at when the current flowing through the access region is small. As the current increases, the access region resistance increases and starts impacting the I-V. This behavior can be seen from Fig. 5.26.

Chapter 6
Trapping Models

6.1 Introduction

The defects during the growth process of GaN HEMTs and the ones arising during the device operation under application of electrical stress cause electrically active trap centers in the device. These trap centers can be present in different locations in the device such as buffer layer, AlGaN/GaN interface layer, and barrier layer. The amount and location of traps can be different from manufacturer to manufacturer, and it can also vary depending on device architecture. For instance, the field-plate design and the passivation layer design are known to affect the trapping behavior.

Several researchers have and continue to investigate the mechanism of the charge trapping [50–54]. Techniques such as luminescence and pulsed I-V characterization have been used to understand the trapping effects. Charge trapping is topic of intense research at the time of writing this book still. Nevertheless, some researchers have proposed gate-leakage and impact ionization as possible two mechanisms via which electrons can get trapped in the trapping centers.

These traps affect device characteristics significantly and give rise to the effects known as drain-lag, gate-lag, knee walk-out, dynamic on-resistance, etc. The compact model development for trapping effects is equally complex issue. First, as the exact physical mechanism is unclear and/or changes with device architecture and technology, model development has to reply on empirical evidence seen in measurements. Second, the requirements of a model capturing the charge-trapping phenomenon are dependent on the end use of the model. As a result, the type and degree of approximations made to develop models of trapping effects also changed from solution to solution.

In ASM-HEMT, multiple models for trapping effects have been introduced. These different modeling approaches can be enabled or disabled by the model configuration switch TRAPMOD. The different manners in each of the TRAPMOD settings use different approximations for modeling the trapping effects. These different approaches have been found suitable for used needs and device

S. Khandelwal, *Advanced SPICE Model for GaN HEMTs (ASM-HEMT)*,
https://doi.org/10.1007/978-3-030-77730-2_6

choice as discussed below. Advanced modeling approach for trapping based on recombination-generation device physics is also discussed in later sections. This advanced modeling approach is not implemented in the open version of the ASM-HEMT model at the time of writing this book.

6.2 Model with Symmetric Time-Constants

Charge trapping by nature is a transient phenomenon. There is a finite needed for trapping of the charges, and similarly for detrapping from the trap centers. It is found that the time scales for the trapping and detrapping can be quite different. However, by modeling the times involved in trapping and detrapping to be symmetric or the same, the model can be relatively simplified. This simple model is still quite useful because for continuous device operation the effective trapped state of the device needs to be modeled. To this end there are two modeling approaches implemented in ASM-HEMT model. They are enabled by setting TRAPMOD = 2 and TRAPMOD = 3.

The topology for TRAPMOD = 2 is shown below. In this case two different RC sub-circuits are used (Fig. 6.1).

The trap voltages V_{trap1} and V_{trap2} update the parameters cut-off voltage, sub-threshold slope, source, and drain resistances to capture the effects of traps. In this case, since the trap generation current is both a function of V_d and V_g this model is more flexible as compared to TRAPMOD = 3 and TRAPMOD = 1 as it allows independent model tuning. Sample modeling result achieved with TRAPMOD = 2 is shown in Fig. 6.2.

The topology for trap-model TRAPMOD = 3 is shown in Fig. 6.3. This is recommended for modeling the GaN power device dynamic ON-resistance. Only the drain-side resistance is affected in this trap model. Example modeling results achieved with TRAPMOD = 3 for power GaN device is shown in Fig. 6.4. This shows the degradation on-resistance modeled under different off-state voltage stress.

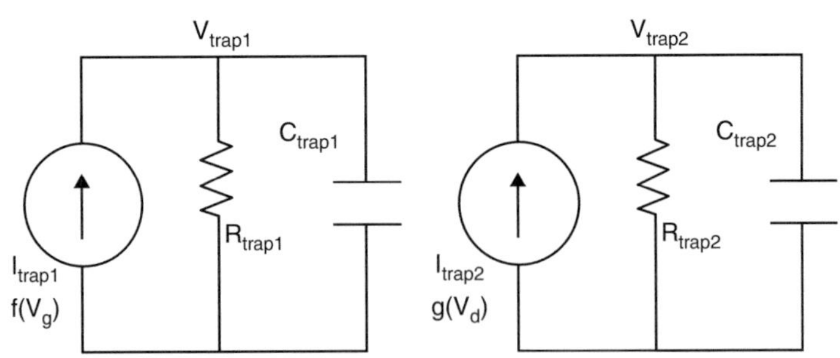

Fig. 6.1 Modeling trapping effects in GaN HEMTs with the voltage Vtrap1 and Vtrap2 in this sub-circuit

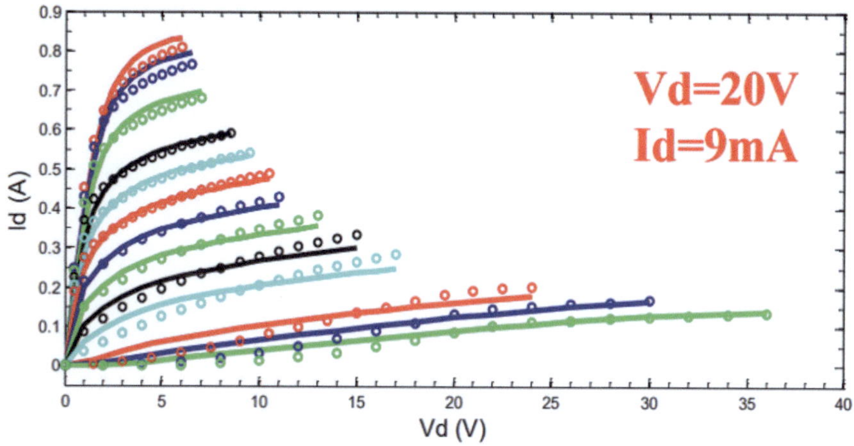

Fig. 6.2 Example modeling results for pulsed I-V data with TRAPMOD = 2

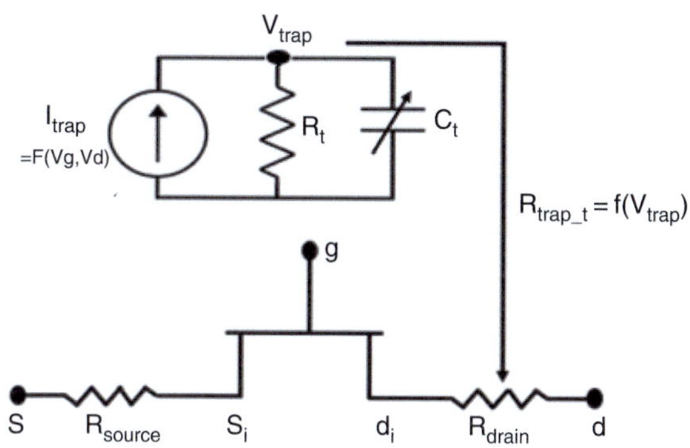

Fig. 6.3 Modeling trapping effects in GaN HEMTs with the RC sub-circuit

6.3 Model with Asymmetric Time-Constants

The asymmetry between the trapping and the detrapping processes is modeled in TRAPMOD = 1. The topology for TRAPMOD = 1 is shown in the Fig. 6.5. In this case, a diode is used to model the asymmetric time-constants. During the trapping process, the diode is forward biased and the capacitor gets charged through the diode. For detrapping, the capacitor gets discharged through the resistor as the diode gets reverse biased. This achieves the asymmetric during charging and discharging events.

The trap-voltage V_{trap} generated is fed back into the model and it updated parameters like the cut-off voltage, DIBL, source, and drain resistances to capture the effects of traps. Sample model results achieved with TRAPMOD = 1 are shown in Fig. 6.6.

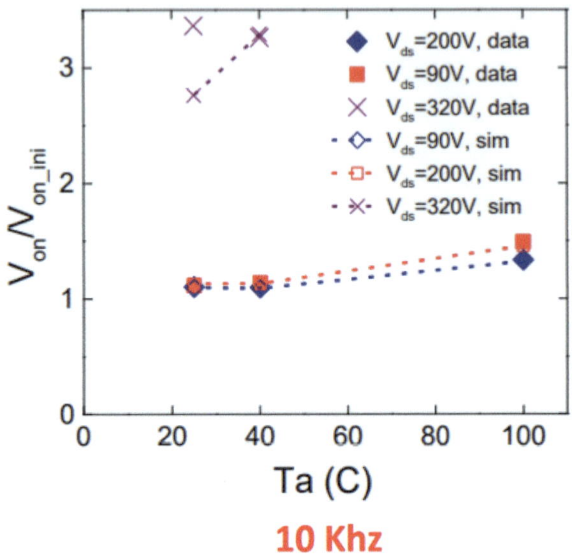

Fig. 6.4 Example modeling results for pulsed I-V data with TRAPMOD = 2

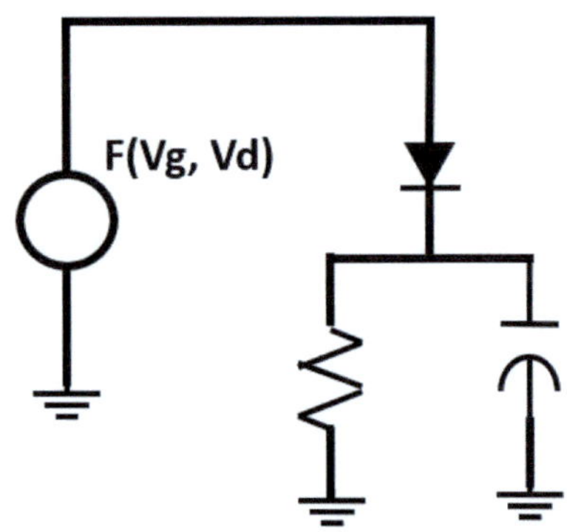

Fig. 6.5 Modeling trapping effects in GaN HEMTs with the voltage Vtrap in this sub-circuit for TRAPMOD = 1

6.4 Trapping Model Based on Recombination Generation

The third approach is based on the SRH-based trap model presented in [55, 56], and unify it with ASM GaN model, which is described in part I of this article [57], to develop a complete model for GaN devices. Unlike the approach presented in [55, 56], where the potential introduced by the SRH-based trap model modifies, depending on the location of the trap centers in the device, the gate voltage and/or the drain and source access resistances, in the GaN-HEMT model presented in this work, this potential modifies, depending on the location of the trap centers in the device, different physical parameters of ASM GaN model.

The developed model is extensively validated by modeling DC and pulsed I-V characteristics of a GaN device. The accuracy of the model is validated for eight different quiescent bias points, with quiescent drain voltage ranging from 5 V to

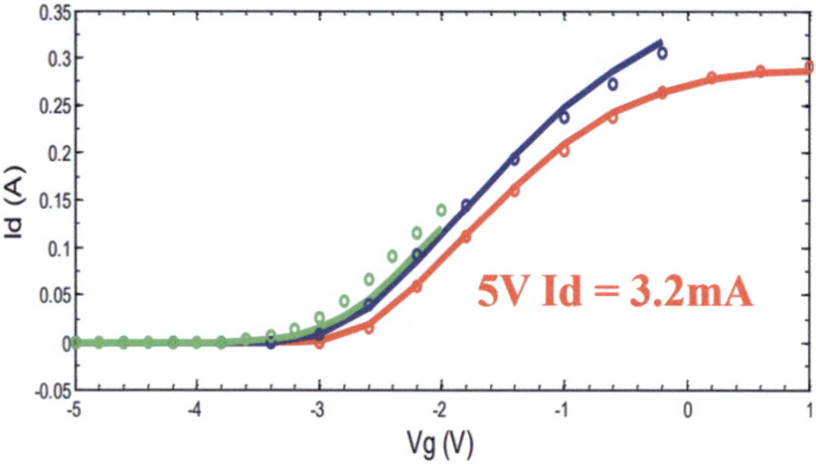

Fig. 6.6 Modeling trapping effects in GaN HEMTs with the voltage Vtrap in this sub-circuit for TRAPMOD = 1

20 V and quiescent gate voltage ranging from −2.8 V to −3.8 V, and a large range of gate and drain voltages to which the device was pulsed in the pulse measurements and at which the device was measured in the DC measurements, with gate voltage ranging from −4 V to 0.4 V and drain voltage ranging from 0 V to 40 V.

The paper is arranged as follows. Section 6.5 presents the results of DC measurements as well as pulsed measurements from eight different quiescent conditions performed on a GaN HEMT. Section 6.6 reviews the charge-trapping mechanism in FETs and the SRH-based trap model developed in [55, 56]. Section 6.7 presents a comprehensive model for the device under test which combines the ASM model with the SRH trap model, yielding an excellent fit to both the DC and pulse measurements. It is shown that, for the device under test, the parameters which are affected by the trap model are VOFF (the cut-off voltage), NS0ACCD (the 2-DEG Charge Density in per square meter in drain access region), and UA (the first-order mobility degradation parameter).

6.5 Measurement Results

DC and pulse measurements were performed on a commercially available GaN HEMT of total 20 μm channel width and 0.25 μm channel length. The pulse width and duration in pulse measurements were 20 μs and 1 ms, respectively. Measurements were performed 10 μm after step change. The DC output and transfer

Fig. 6.7 Measured (symbol) and modeled (solid line) DC output characteristics (**a**) and DC transfer characteristics (**b**) of the device under test. Self-heating was accounted for using the standard thermal network

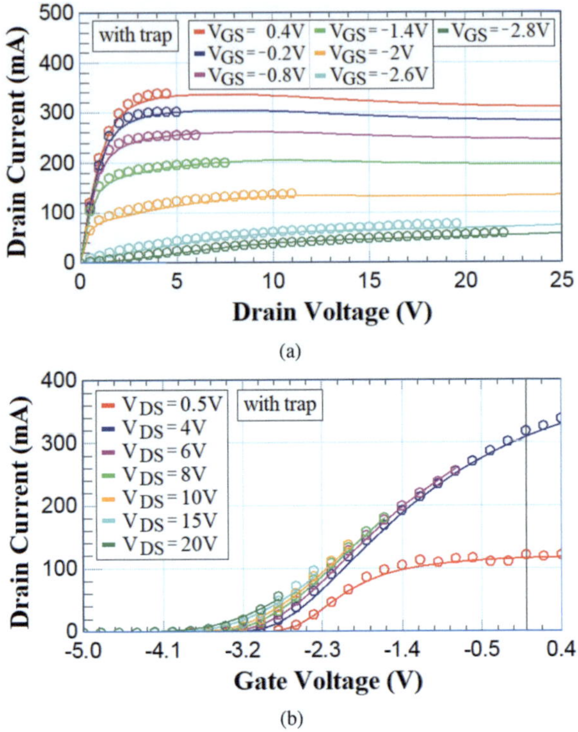

characteristics of this device are shown in Fig. 6.7. The device was pulsed from the following eight different quiescent conditions:

- $(V_{DSQ}, V_{GSQ}) = (5\text{ V}, -3.2\text{ V}), (5\text{ V}, -2.8\text{ V})$
- $(V_{DSQ}, V_{GSQ}) = (10\text{ V}, -3.4\text{ V}), (10\text{ V}, -2.9\text{ V})$
- $(V_{DSQ}, V_{GSQ}) = (15\text{ V}, -3.6\text{ V}), (15\text{ V}, -3.0\text{ V})$
- $(V_{DSQ}, V_{GSQ}) = (20\text{ V}, -3.8\text{ V}), (20\text{ V}, -3.1\text{ V})$

to a set of final bias points, with final drain voltage ranging from 0 V to 40 V with a step size of 0.5 V and final gate voltage ranging from −4 V to 0.4 V with a step size of 0.2 (Fig. 6.8).

Measurement results indicate that the I-V characteristics of the device exhibit, as shown in Fig. 6.9, significant variation with the quiescent drain voltage, but are insensitive, as shown in Fig. 6.10, to the quiescent gate voltage. From the results of Fig. 6.9, one can note the following:

- There is a positive shift in the threshold voltage with increasing V_{DSQ}.
- There is a decrease in the on-state current with increasing V_{DSQ}.

These observations suggest that to model the trapping behavior of the device under test, one must consider an acceptor-like trap center which modifies the parameter VOFF and probably the parameter NS0ACCD of the ASM model. Indeed, as we

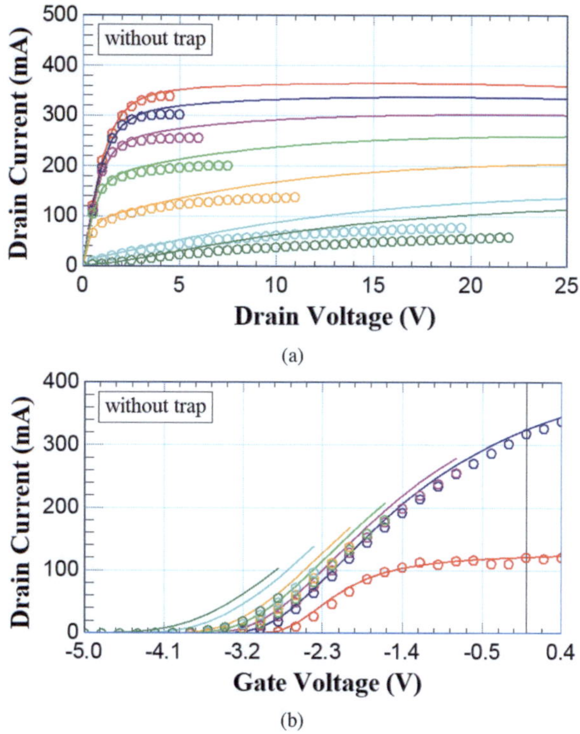

Fig. 6.8 The simulation results of Fig. 6.7 for when the trap model is disabled (**a**) Output characteristics. (**b**) Transfer characteristics

will show later, it is possible to obtain an excellent fit to the output and transfer characteristics of the device by letting the trap center of the device to change these two parameters, as well as the parameter UA.

Figures 6.11 and 6.12 show the full sets of output and transfer characteristics corresponding to the four different quiescent drain voltage (i.e., $V_{DSQ} = 5$ V, 10 V, 15 V, and 20 V).

After reviewing the circuit model developed in [55, 56] we will show that the observed complicated trapping behavior in the measured results can be modeled by combining the ASM model with an SRH-based model of an acceptor trap center which behaves as an electron trap center in the region of operation of the device and is subject to an electron current which is composed of the electrons of the channel which get injected into the trap center through tunneling. The potential of this trap center, which represents the ionization density of the trap center, modifies three parameters of the ASM model, namely VOFF, UA, and NS0ACCD.

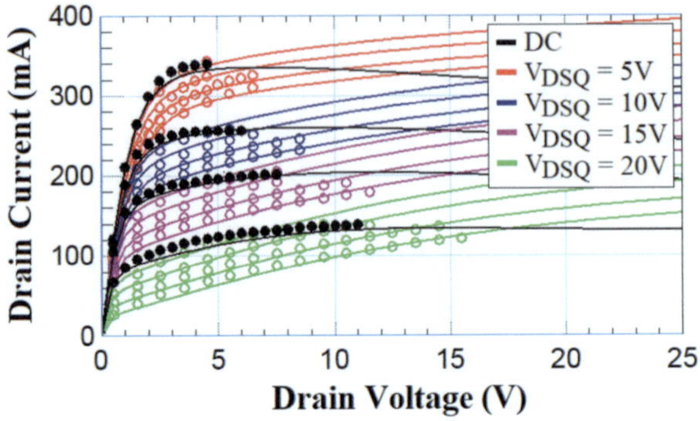

(a)

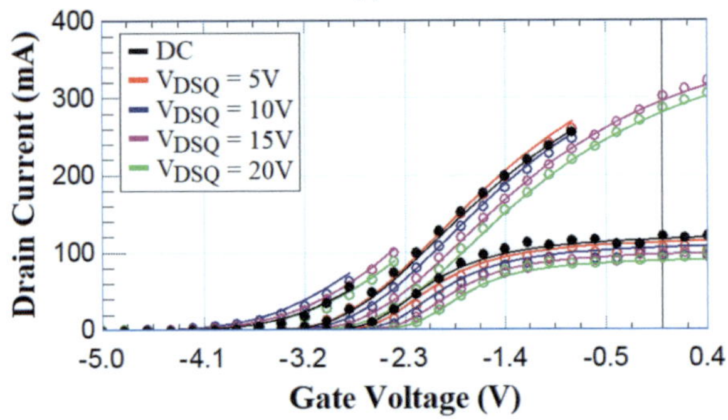

(b)

Fig. 6.9 Measured (symbol) and modeled (solid line) pulsed output characteristics (**a**) and transfer characteristics (**b**) of the device under test for the following quiescent bias points: (V_{GSQ}, V_{DSQ}) $=(5\,\text{V}\,, -3.2\,\text{V})$, $(10\,\text{V}\,, -3.4\,\text{V})$, $(15\,\text{V}\,, -3.6\,\text{V})$, and $(10\,\text{V}\,, -3.8\,\text{V})$. The gate voltages for the output characteristics are: 0 V, −0.8 V, −1.4 V, and −2.0 V. The drain voltages for the transfer characteristics are: 0.5 V, 6 V, and 20 V. The output and transfer characteristics are shown, for the purpose of comparison, alongside with the DC characteristics of Fig. 6.11 for the same gate voltages

6.6 Circuit Implementation of a Trap Center

In this section we review the charge-trapping mechanism in FETs and the SRH-based trap model developed in [55, 56].

6.6.1 Rates of Charge Capture and Emission at a Trap Center

There exists four competing processes for hole and electron capture and emission at the energy level E_x of a trap center. These processes are electron capture, electron emission, hole capture, and hole emission. In a Shockley–Read–Hall (SRH)

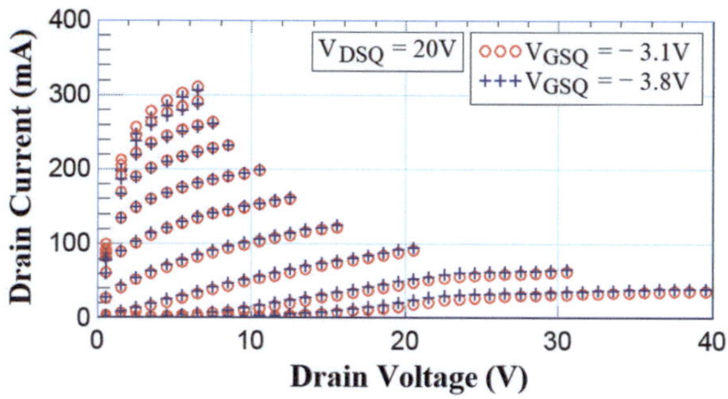

Fig. 6.10 Measured pulsed output characteristics of the device under test for the quiescent bias point indicated in the legend. Gate voltage ranges from −3.2 V to 0.4 with a step size of 0.4. This figure shows that the pulsed output characteristic of the device exhibits negligible dependence on the quiescent gate voltage. Similar results were obtained for other quiescent bias points

trapping process [58], which is the concern of this paper, these processes are governed by Fermi–Dirac statistics. There is a rate associated with each process. These rates are defined, respectively, as the rate of electron capture from the conduction band, R_{cn}, the rate of electron emission to the conduction band, R_{en}, the rate of hole capture from the valence band, R_{cp}, and the rate of hole emission to the valence band, R_{ep}.

Let $n_x\left[\mathrm{m}^{-3}\right]$ and $p_x\left[\mathrm{m}^{-3}\right]$ denote, respectively, the concentrations of the captured electrons and holes at a trap center. For a unit volume of material, the net rate of electron accumulation at the energy levels of a trap center is given by

$$dn_x/dt = U_n - U_p, \tag{6.1}$$

where $U_n \equiv R_{cn} - R_{en}$ and $U_p \equiv R_{cp} - R_{ep}$ are, respectively, the total rate of electron and hole capture at E_x and are given by [58]

$$U_n = e_n\left[p_x \exp\left(\phi_n\right) - n_x\right] \text{ and} \tag{6.2}$$

$$U_p = e_p\left[n_x \exp\left(\phi_p\right) - p_x\right], \tag{6.3}$$

where $e_n\left[\mathrm{s}^{-1}\right]$ and $e_p\left[\mathrm{s}^{-1}\right]$ are, respectively, the characteristic frequencies of electron and hole emissions, and

$$\phi_n \equiv \left(F_n - E_x\right)/kT \text{ and} \tag{6.4}$$

$$\phi_p \equiv \left(E_x - F_p\right)/kT, \tag{6.5}$$

where F_n [eV] and F_p [eV] are, respectively, the quasi-Fermi levels for electrons and holes, k [J/K] is the Boltzmann constant, and T [K] is temperature.

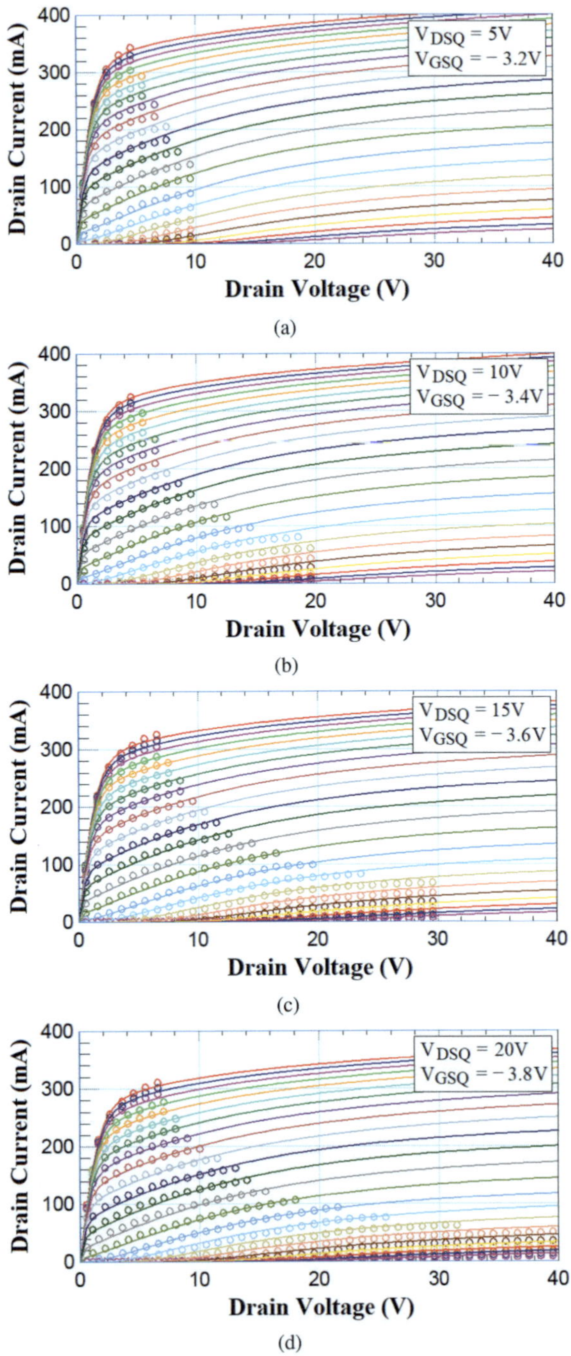

Fig. 6.11 Measured (symbol) and modeled (solid line) pulsed output characteristics for the various quiescent bias points indicated in the legends. The gate voltage ranges from −3.2 V to 0.4 V with a step size of 0.2 V (**a**) VDSQ = 5 V and VGSQ = −3.2 V. (**b**) VDSQ = 10 V and VGSQ = −3.4 V. (**c**) VDSQ = 15 V and VGSQ = −3.6 V. (**d**) VDSQ = 20 V and VGSQ = −3.8 V

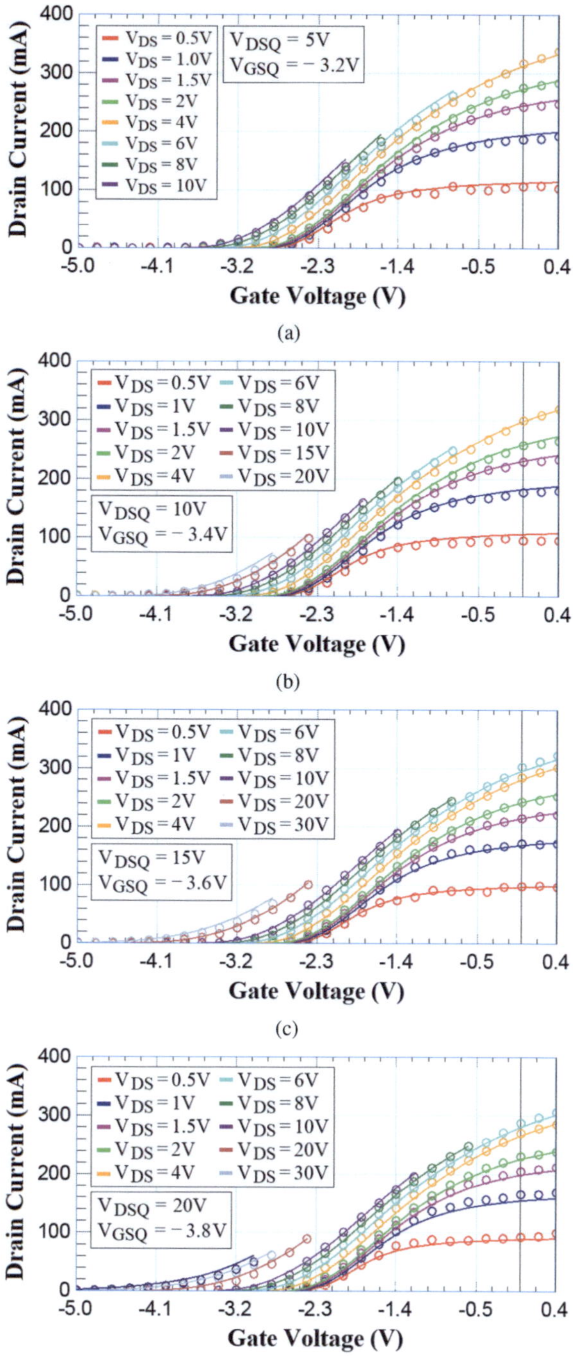

Fig. 6.12 Measured (symbol) and modeled (solid line) pulsed transfer characteristics for the various quiescent bias points indicated in the legends. The drain voltages are also indicated in the legends

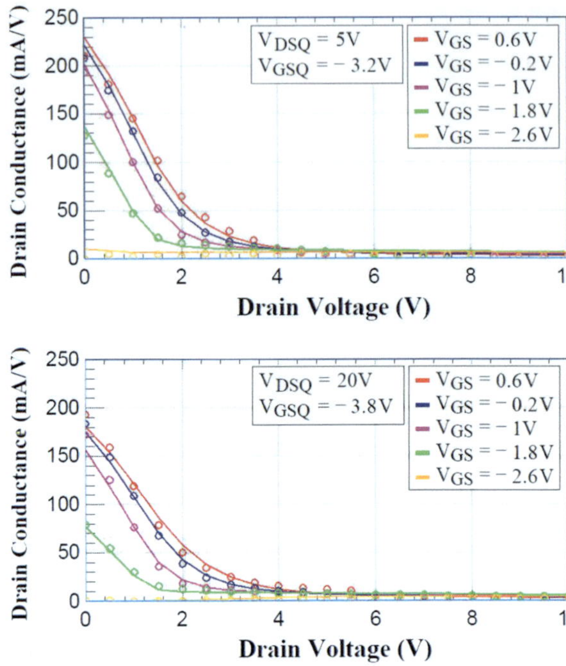

Fig. 6.13 Measured (symbol) and modeled (solid line) pulsed output characteristics for the two quiescent bias points indicated in the legends. The gate voltage ranges from −3.2 V to 0.4 V with a step size of 0.2 V

If the trap center is an acceptor,[1] then (6.1) corresponds to the ionization rate of the trap center, and if it is a donor,[2] then (6.1) corresponds to the de-ionization rate of the trap center (Figs. 6.13 and 6.14).

6.6.2 *Control Potential of a Trap Center*

The concentrations of electrons and holes available for capture at a trap center, denoted respectively as $n\ \left[\mathrm{m}^{-3}\right]$ and $p\ \left[\mathrm{m}^{-3}\right]$, are given by [59]

$$n = \hat{n}\ \exp\left(\phi_n\right)\ \text{ and} \tag{6.6}$$

$$p = \hat{p}\ \exp\left(\phi_p\right), \tag{6.7}$$

[1] An acceptor trap center is a trap center which is neutral when all the energy levels of the trap center are empty, and negatively charged and, therefore, ionized, when one or more energy levels of the trap center are occupied by electrons.

[2] A donor trap center is a trap center that is neutral when all the energy levels of the trap center are occupied by electrons, and positively charged and, therefore, ionized, when one or more energy levels of the trap center are empty.

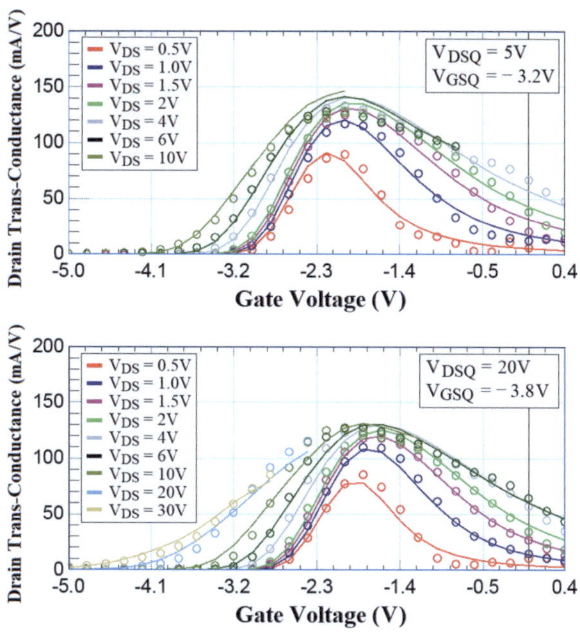

Fig. 6.14 Measured (symbol) and modeled (solid line) pulsed transfer characteristics for the two quiescent bias points indicated in the legends. The drain voltages are also indicated in the legends

where $\hat{n}\ \left[\mathrm{m}^{-3}\right]$ is the concentration of the electrons in the conduction band when F_n falls at E_x, and $\hat{p}\ \left[\mathrm{m}^{-3}\right]$ is the concentration of the holes in the valence band when F_p falls at E_x. From (6.2) and (6.3), ϕ_n and ϕ_p act as controlling potentials for the net rate of electron accumulation at a trap center, normalized to the thermal energy kT. From (6.6) and (6.7), ϕ_n and ϕ_p act as controlling potentials for the available charge for capture at a trap center. This leads to an interpretation of the terms ϕ_n and ϕ_p as the normalized control potentials of a trap center to the thermal energy (in eV). For simplicity, ϕ_n and ϕ_p will be referred to as the control potentials of a trap center.

6.6.3 Current Injection into a Trap-Center Region

The injection of electrons and holes into a trap center modifies the Fermi levels F_n and F_p at the trap center and, consequently, modifies the control potentials ϕ_n and ϕ_p of the trap center. Let $n_o\ \left[\mathrm{m}^{-3}\right]$ and $p_o\ \left[\mathrm{m}^{-3}\right]$ be, respectively, the densities of the electrons and holes available for capture at a trap center before carrier injection, and n_J and p_J be, respectively, the densities of the injected electrons and holes in the trap-center region. The densities of the electrons and holes available for capture at a trap center after carrier injection is then given by $n = n_o + n_\mathrm{J}$ and $p = p_o + p_\mathrm{J}$. Let I_n [A] and I_p [A] be, respectively, the currents of injected electrons and holes into the trap center. The concentrations n_J and p_J are linearly related to the currents I_n

and I_p.[3] From (6.6) and (6.7), the control potentials of a trap center in the presence of carrier injection are given by

$$\phi_n = \ln\left(n_o/\hat{n} + \alpha_n\, I_n \right) \text{ and} \tag{6.8}$$

$$\phi_p = \ln\left(p_o/\hat{p} + \alpha_p\, I_p \right), \tag{6.9}$$

where α_n and α_p are constants. In FETs, the terms n_o, p_o, I_n, and I_p are all functions of the gate–source and gate–drain terminal potentials, V_{GS} and V_{DS}, of the FET.

6.6.4 Circuit Model of a Trap Center

Depending on the bias condition, a trap center can either act as an electron or hole trapping center or as a carrier generation or recombination center. The criteria for electron trapping, hole trapping, carrier recombination, and carrier generation are given, respectively, by [59]

$$R_{cp} \ll R_{en} \text{ and } R_{cn} \gg R_{ep}; \tag{6.10}$$

$$R_{cp} \gg R_{en} \text{ and } R_{cn} \ll R_{ep}; \tag{6.11}$$

$$R_{cp} \gg R_{en} \text{ and } R_{cn} \gg R_{ep}; \text{ and} \tag{6.12}$$

$$R_{cp} \ll R_{en} \text{ and } R_{cn} \ll R_{ep}. \tag{6.13}$$

In general, depending on the bias condition, a trap center, regardless of its polarity, being an acceptor or a donor, acts either as an electron or hole trapping center or as a generation or recombination center. The following describes the comprehensive model of a trap center in an FET developed in [56], suitable for circuit simulators, that can capture the trapping behavior at various bias conditions.

A trap capacity per unit volume $C_X\left[\mathrm{F\,m^{-3}}\right]$ can be defined in terms of a trapping potential v_X [V] and the concentration of the electrons n_X in the trap center as follows:

$$C_X = (-q)\, n_X \,/\, v_X, \tag{6.14}$$

where q [C] is the electron charge. The dynamics of the charge density n_X can be described, substituting (6.14) in (6.1), by a current source i_X [A] driving the capacitor C_X to a potential difference of v_X:

$$i_X = (i_{cn} - i_{en}) - (i_{cp} - i_{ep}), \tag{6.15}$$

[3] Consider, for instance, a sheet of trap centers. Let A_X [m^2] be the cross-sectional area perpendicular to the direction of the "injected" current. The currents I_n and I_p are related to n_J and p_J by $I_n = (v_n\, q\, A_X)\, n_J$ and $I_p = \left(v_p\, q\, A_X\right)\, p_J$, where v_n [m/s] and v_p [m/s] are the drift velocities of the injected electrons and holes, respectively.

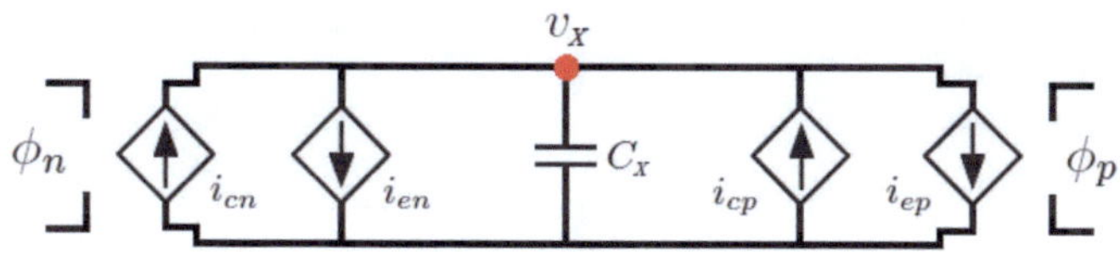

Fig. 6.15 Circuit model for a trap center

where $i_{cn} \equiv (-q)\, R_{cn}$ and $i_{en} \equiv (-q)\, R_{en}$ define electron-capture and electron-emission currents, respectively, and $i_{cp} \equiv (-q)\, R_{cp}$ and $i_{ep} \equiv (-q)\, R_{ep}$ define hole-capture and hole-emission currents respectively. These currents are given by

$$i_{cn} = C_x\, e_n\, (V_x - v_x)\, \exp(\phi_n)\,, \tag{6.16}$$

$$i_{en} = C_x\, e_n\, v_x\,, \tag{6.17}$$

$$i_{cp} = C_x\, e_p\, v_x\, \exp\left(\phi_p\right) \text{ and} \tag{6.18}$$

$$i_{ep} = C_x\, e_p\, (V_x - v_x)\,, \tag{6.19}$$

where $V_x \equiv (-q\, N_x)\,/C_x < 0$ is the potential v_x when the energy levels of the trap center are fully occupied by electrons, where $N_x\left[\mathrm{m}^{-3}\right]$ is the number per unit volume of the energy levels E_x at a trap center, and is such that $N_x = n_x + p_x$. If the trap center is an acceptor, then its trapping potential v_x is directly proportional to its ionization density n_x, whereas if the trap center is a donor, then, using (6.14) and the relation $n_x + p_x = N_x$, the potential $v_x - V_x$ is directly proportional to its ionization density p_x. The circuit model is shown in Fig. 6.15.

6.6.5 *Steady-State Response*

In a steady-state condition, the total emission and capture current i_x (6.15) into a trap center is zero. This leads to the following relation between the trapping potential v_x at a steady-state condition, denoted as $\overline{v}_x$, and the control potentials ϕ_n and ϕ_p:

$$\overline{v}_x = \frac{\left[1 + \exp(\phi_n + \epsilon)\right]}{\left[1 + \exp(\phi_n + \epsilon)\right] + \eta\left[1 + \exp\left(\phi_p - \epsilon\right)\right]}\, V_x\,, \tag{6.20}$$

where $\eta \equiv e_n/e_p$ and $\epsilon \equiv \ln(\eta)$.

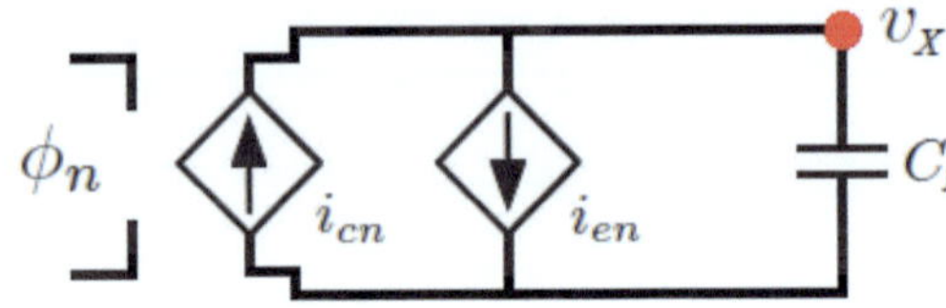

Fig. 6.16 Circuit model for a trap center acting as an electron-trapping center

6.6.6 *Circuit Model for an Electron-Trapping Center*

The measurement results of Fig. 6.9 suggest that the observed dispersion in the I-V characteristics of the device under test is caused by a trapping process at an acceptor-like trap center. Trapping was observed for quiescent bias points at which the device was on, suggesting that the trapping process is triggered by the electrons of the channel which are injected into the trap center through tunneling. It is plausible, therefore, to assume that in the region of operation of the device, the trap center is behaving as an electron trap center. This assumption is further justified by the reasonable fit we obtained to the measured data by modeling the observed dispersion assuming that, in the region of operation of the device, the trap center is an electron-trapping center.

When the condition of electron trapping (6.10) is satisfied, (6.20) simplifies to

$$\overline{v}_X = V_X / \left[1 + \exp\left(-\phi_n\right)\right]. \tag{6.21}$$

We can reach this equation by setting the parameter e_p (the characteristic frequency of hole emission) of the trap center to zero. This then simplifies the circuit model of Fig. 6.15 to the one shown in Fig. 6.16.

6.7 Discussion

It is possible to model the observed complicated trapping behavior in the measured results of Sect. 6.5 by considering a trap center with a steady-state potential of the form (Fig. 6.17)

$$\overline{v}_X = \frac{V_X}{1 + \dfrac{1}{\exp\left(A\,V_{GS} + B\,V_{DS} + C\right) + D}}, \tag{6.22}$$

where $A = 0$, $B = 0.25$, $C = -3$, and $D = 0.15$ and letting the trap potential modify the three parameters VOFF, UA, and NS0ACCD of the ASM model in a linear manner as:

$$\text{VOFF}_X = \text{VOFF} + \kappa_1\,\tilde{v}_X, \tag{6.23}$$

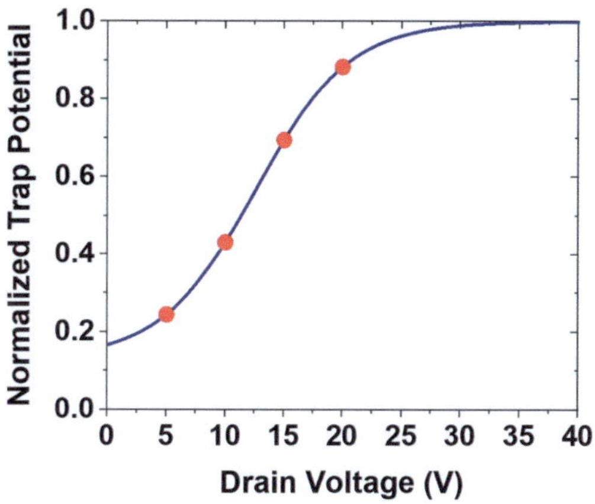

Fig. 6.17 The steady potential of the trap center in the device under, normalized to V_X, which is the potential of the trap center when it is fully ionized. The red symbols correspond to the trap potentials at the quiescent bias conditions at which the device under test was held in the pulse measurements. The blue line shows the trap potential at all the drain voltages, ranging from 0 V to 40 V, at which the DC characteristics of the device were measured. The trap potential at zero drain voltage is the initial potential of the trap center in the absence of current injection into the trap center, which is then modified by the current injected into the trap center after turning on the device. The parameter $D \equiv n_o/\hat{n}$ in (6.22) controls the initial state of the trap center

$$\mathrm{UA}_x = \mathrm{UA} + \kappa_2\, \tilde{v}_x \quad \text{and} \tag{6.24}$$

$$\mathrm{NS0ACCD}_x = \mathrm{NS0ACCD} + \kappa_3\, \tilde{v}_x, \tag{6.25}$$

where $\tilde{v}_x = v_x/V_x$ is the normalized trap potential, VOFF $= -2.6$, $\kappa_1 = 0.54$, UA $= 120 \times 10^{-9}$, $\kappa_2 = 23 \times 10^{-9}$, NS0ACCD $= 1.7 \times 10^{17}$, and $\kappa_3 = -9 \times 10^{16}$. This implies that first-order modification to VOFF, UA, and NS0ACCD is sufficient to capture the effect of trapping. The variations of VOFF_x, UA_x, and $\mathrm{NS0ACCD}_x$ with drain voltage are illustrated in Fig. 6.18.

In a pulse measurement, the trap potential that modifies the electrical performance of the device is a single potential given by (6.22) where V_{GS} and V_{DS} are the quiescent gate and drain potentials. In a DC measurement, the trap potential which modifies the electrical performance of the device varies with the bias point. The simulation results with the parameters listed above are shown in Figs. 6.7, 6.9, 6.11, 6.12, 6.13, 6.14, and 6.15. Also, high frequency (10 GHz) large signal RF validation of the model for a quiescent operating point of $(V_{DS}, V_{GS}) = (15\,\text{V}, -3.6\,\text{V})$ and for optimal complex load condition is shown in Fig. 6.19.

The above steady-state form for the trap potential, which is illustrated in Fig. 6.17, can be explained theoretically in the following manner. There exists an acceptor-like trap center in the device under test which, in the region of operation of the device, behaves as an electron trap center. The current injected into the trap center is the current composed of the electrons of the channel which are

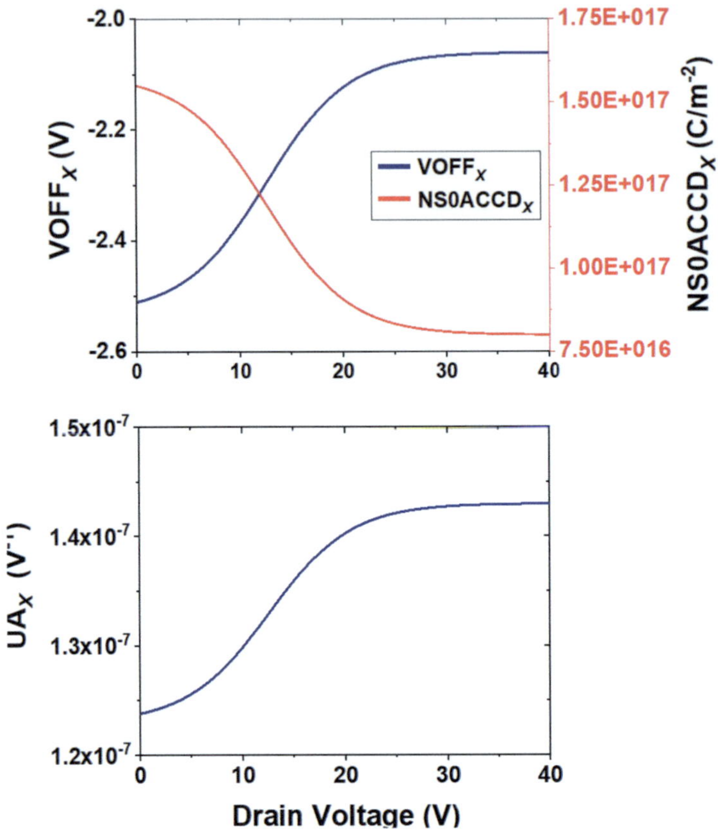

Fig. 6.18 The variations of VOFF_X, UA_X and $\mathrm{NS0ACCD}_X$, described by (6.23), (6.24), and (6.25) respectively, with drain voltage

injected toward the trap center by a tunneling process, and so have an exponential dependence on the terminal potentials:

$$I_n = I_o \exp\left(A\, V_{GS} + B\, V_{DS}\right), \tag{6.26}$$

where I_o, A, and B are constants. The measurement results show negligible dependence on the gate–source potential, so we can set $A = 0$. The control potential ϕ_n, given by (6.8), becomes, using (6.26),

$$\phi_n = \ln\left[\exp\left(B\, V_{DS} + C\right) + D\right], \tag{6.27}$$

where $C \equiv \ln(\alpha_n I_o)$ and $D \equiv n_o/\hat{n}$. Substituting (6.27) in (6.21) we recover (6.22).

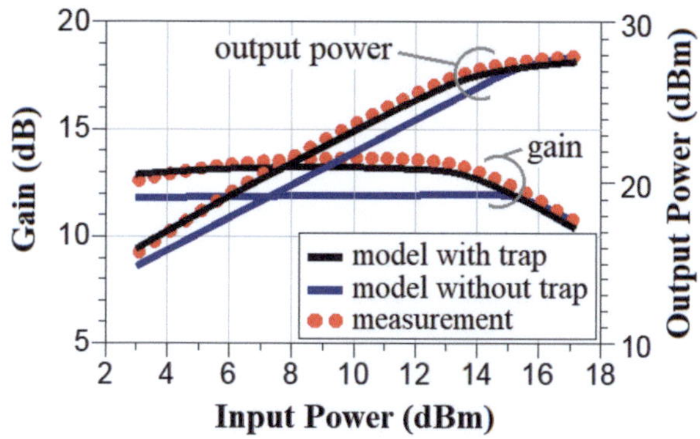

Fig. 6.19 High frequency (10 GHz) large signal RF validation of the model presented in this work for a quiescent operating point of (V_{DS}, V_{GS}) = (15 V, −3.6 V) and optimal complex load condition

To find the parameters of the ASM model which are affected by the trap potential we took the following approach. We considered the output and transfer characteristics for one of the quiescent bias points (in particular (V_{GSQ}, V_{DSQ}) = (20 V, −3.8 V)) as a reference set of characteristics. Self-heating in the ASM model was turned off, and the parameters for the ASM model were optimized to obtain a reasonable fit to the reference output and transfer characteristics. The measurement results of Fig. 6.9 clearly show that there is a shift in the threshold voltage with quiescent drain voltage, so we began by tuning VOFF of the model to obtain reasonable fits to the output and transfer characteristics for the other quiescent conditions. However, we found that to obtain the excellent fits shown in Figs. 6.9, 6.11, and 6.12, NS0ACCD and UA must also be tuned. We then turned on the self-heating process in the ASM model and fine tuned some of the parameters of the ASM model for a global optimization for the set of parameters which give excellent fits to both DC, shown in Fig. 6.7, and pulsed output and transfer characteristics.

We extracted the parameters of the trap model in the following manner. We plotted the parameters VOFF, UA, and NS0ACCD versus quiescent bias point, and by fitting to these curves an equation of the form given by (6.22) we extracted the parameters A, B, C of (6.22) and the parameters κ_1, κ_2, and κ_3 of (6.23), (6.24), and (6.25).

Chapter 7
Non-Ideal Effects in GaN Capacitances and Their Modeling

7.1 Introduction

Capacitance in a transistor can be modeled with many different approaches. In earlier models [60], explicit formulations for modeling device capacitances were directly used. Later it was found that this approach does not conserve charge and can lead to non-convergence in simulations [61]. As a result, charge based models for capacitive effects were developed. These models naturally conserve charge. In ASM-HEMT model, this approach is used to model the device capacitances. The terminal charges of the device are calculated based on device physics. These formulations were discussed in Chap. 4. The derivative of terminal charges w.r.t specific terminal voltage models the terminal capacitance. For instance, gate–drain capacitance C_{gd} is given by taking derivative of gate charge Q_g w.r.t drain-voltage V_d.

$$C_{gd} = -\frac{\partial Q_g}{\partial V_d}. \tag{7.1}$$

And gate–source capacitance is given by taking derivative of gate charge Q_g w.r.t source-voltage V_s.

$$C_{gd} = -\frac{\partial Q_g}{\partial V_s}. \tag{7.2}$$

Other device capacitances can be derived in similar manner. As the charge formulations account for the law of charge conservation, this technique of capacitance models naturally preserves it. Additionally, the multi-bias dependence of all the device capacitances is modeled with terminal charge formulations. For instance, C_{gs} and C_{gd} have both drain–source V_{ds} and gate–source V_{gs} dependence. The V_{ds}

S. Khandelwal, *Advanced SPICE Model for GaN HEMTs (ASM-HEMT)*,
https://doi.org/10.1007/978-3-030-77730-2_7

and V_{gs} dependence of these capacitances is captured by one analytical model for gate charge Q_g.

Another important aspect of charge or capacitance modeling is the consistency with DC current formulations. The consistency of these two aspects of device behavior is crucial for model change in device behavior with process technology. For instance, if cut-off voltage changes in device due to a fabrication process change, it will impact both current and the capacitance of the device. If these two aspects of device behavior are modeled independent from each other, the correlation between capacitance and current is lost in the model. Similarly, a change in gate-length due to manufacturing variations will lead to a change in both current and capacitance. An increase in gate-length will increase total capacitance, while it will decrease the current. A consistent model accounts for this behavior.

The capacitance simulated from the model is dependent on derivative of terminal charges. The DC current has formulations independent of the terminal charge formulations. However, for a consistent model it is necessary that these two formulations should come from same underlying physics of the device. The device electrostatics causing gate control of the drain current also is the underlying reason for increase in capacitance with gate voltage. This consistency in charge and current formulations is preserved in ASM-HEMT model as both the current and capacitance formulations use the same surface-potential. Charge and current formulations share the ASM-HEMT parameters. This is how the consistency between charge and current formulations is preserved.

The core charge formulations derived in Chap. 4 consider the ideal device. In GaN HEMTs several real device effects impact the capacitances. These are modeled in ASM-HEMT with same strategy as taken in the current formulations accounting for real device effects. The core model formulations are augmented with real device effects with parameters introduced to model various real device effects.

7.2 Impact of Fringe Capacitances

Layout of a device includes metal connections for gate, drain, and source terminals. A passivation dielectric is usually present between the metal contacts. The field lines from one metal contact to the other give rise to capacitive effects which were not included in the core model formulations. These are accounted for the ASM-HEMT model. These capacitances are equivalent to parallel plate capacitance and are independent of applied bias voltage.

The metal contact for gate and the source electrode have fringe electric field between them, giving rise to capacitance. This capacitance is dependent on the distance between the gate and source metal. However, fabricated device is complex, and the metal electrodes are more likely not ideally parallel to each other. To address this, a model parameter CGSO is introduced in the ASM-HEMT model. CGSO is gate–source capacitance when the channel is completely off. To include the effect of CGSO, the total gate charge is obtained by adding charge due to CGSO to the core gate charge. This results in,

$$Q_g = Q_{gi} + CGSO \cdot V_{gs}. \tag{7.3}$$

To get the gate–source capacitance of the device, we take derivative of Eq. (7.3) w.r.t V_{gs},

$$C_{gs} = -\frac{\partial Q_g}{\partial V_s} = -\frac{\partial Q_{gi}}{\partial V_s} + CGSO. \tag{7.4}$$

When the device is off $\frac{\partial Q_{gi}}{\partial V_s}$ is nearly zero and $C_{gs} \approx$ CGSO. So, CGSO is the ASM-HEMT parameter to model the fringe capacitance between gate and source metal electrodes.

The fringe field effects also exist between drain and gate, and drain and source electrodes. These are modeled in similar manner as C_{gs} above. The model parameters for gate–drain and drain–source capacitance are CGDO and CDSO, respectively. The gate charge formulation gets augmented for gate–drain fringe field effect and is given by,

$$Q_g = Q_{gi} + CGSO \cdot V_{gs} + CGDO \cdot V_{ds}. \tag{7.5}$$

For drain–source fringe field effect, the drain charge formulation gets augmented as follows:

$$Q_d = Q_{di} + CDSO \cdot V_{ds}. \tag{7.6}$$

The parameters CGSO, CGDO, and CDSO are crucial in modeling the off-state capacitance of the GaN device.

7.3 Impact of Field-Plates on Device Capacitance

Field-plates are used in GaN HEMTs to tailor the electric field profile and improve the breakdown voltage. The field-plates introduce complex capacitance behavior in GaN HEMTs. The field-plates can be connected to source- or gate-electrode. GaN device can have several field-plates which are connected to gate- or drain-electrode in different configuration.

The 2-DEG charge under the field-plate is influenced by the voltage at the field-plate electrode. As a result, the capacitance introduced by field-plate is dependent on bias voltage. This is a key distinction in the capacitance arising due to the field-plates from the fringe capacitances discussed in previous section. Fringe capacitances do not change with applied bias voltages.

Typically field-plates are farther away from the 2-DEG channel as compared to the main gate electrode. The increased distance from the 2-DEG to the metal electrode over it causes the cut-off voltage of these regions to be more negative as compared to the main gate. The difference in the cut-off voltage of the main gate and the field-plate region and the geometry of these regions cause complex capacitance characteristics. These are modeled in ASM-HEMT model.

To model devices from different technologies and different field-plate configurations ASM-HEMT has a configurable field-plate modeling approach. ASM-HEMT field-plate model allows modeling of total eight field-plates, up to four on each side of the gate. Each field-plate can be configured to be a gate- or source-connected field-plate by setting model parameter FPxMOD where x is the number of the field-plate starting from 1. The sub-sections below discuss the effect of field-plates on capacitances further.

7.3.1 Gate–Source Capacitance

The field-plate between the gate and source electrode of the GaN device impacts gate–source capacitance C_{gs}. The overhang of the T-gate in the gate–source access region also created a gate-connected field-plate. This impacts the behavior of C_{gs} with the variation in gate–source voltage.

The cut-off voltage of the field-plate region is typically lower than the cut-off voltage of the main channel region as the metal electrode is farther away from the 2-DEG charge in this region. This gives rise to complex voltage dependence of C_{gs} as compared to a device without the field-plate. Figure 7.1 shows C_{gs} versus V_{gs} characteristics of a GaN HEMT with and without field-plate. When V_{gs} is very large negative and below the cut-off voltage of the field-plate, 2-DEG in all regions of the device is depleted. As a result, C_{gs} is small and is mainly due to the fringe field capacitance due to the metal lines in the device layout. As V_{gs} is increased the capacitance stays at the low value for the device without field-plate till V_{gs} reaches the cut-off voltage of the main channel region. For the device with the source field-plate, we note the hump-like behavior. This occurs as V_{gs} reaches the cut-off voltage of the source field-plate region. The 2-DEG under the source field-plate is formed as V_{gs} increases beyond the cut-off voltage of this region. This gives rise to the increase in capacitance and the hump-like feature seen in C_{gs} versus V_{gs} characteristics. C_{gs} due to the 2-DEG in the source field-plate region saturates once the 2-DEG is fully formed.

It is interesting to note that the capacitance behavior near the cut-off voltage of the field-plate and the main channel region is also typically different. This is due to the different sub-threshold slopes of these regions. Field-plate regions with their control gate farther away from 2-DEG typically have a poor sub-threshold slope causing the capacitance to rise gradually, than that of the rise in capacitance when

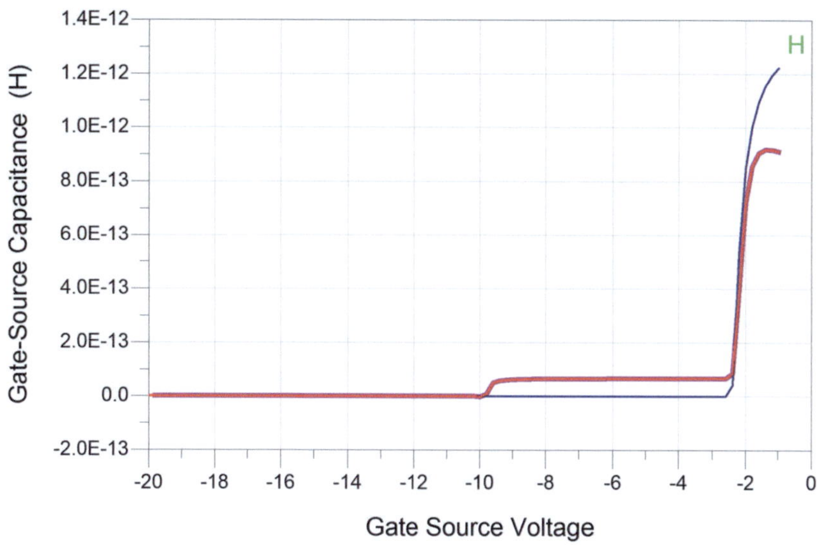

Fig. 7.1 Gate-to-source capacitance versus V_{gs} of a GaN transistor with and without field-plate. V_{gs} is varied from value below the cut-off voltage of the field-plate region

the main channel turns on. This effect is captured with the different sub-threshold parameters for field-plate and main channel regions in the ASM-HEMT model.

7.3.2 Gate–Drain Capacitance

Gate–drain capacitance C_{gd} is very crucial to analyze stability in case of high-speed amplifiers, and the Miller ratio in case of power transistor. C_{gd} is affected by presence of field-plates in the device. In a transistor without field-plate, C_{gd} monotonically reduced with increase in V_{ds} as shown in Fig. 7.2. This is because with an increase of V_{ds} the charge profile in the transistor channel becomes tapered toward drain, making the C_{gd} reduce with V_{ds}. The control of drain voltage over channel charge monotonically decreases with increasing V_{ds}. In the presence of field-plate C_{gd} versus V_{ds} becomes significantly different than usual monotonic decrease. This is due to the different cut-off voltage and geometry of the field-plate region of the device.

In Fig. 7.3 C_{gd} versus V_{ds} for device with gate-field only, and with gate- and source field-plates is shown. In case of device without field-plate C_{gd} monotonically decreases with V_{ds}. In case of device with gate-connected field-plate a hump is observed. This can be understood in the following manner. At a very high V_{ds} the 2-DEG under all regions of the device is depleted. As a result, C_{gd} is small and it is mainly due to the fringe capacitance between gate and drain metal lines of the

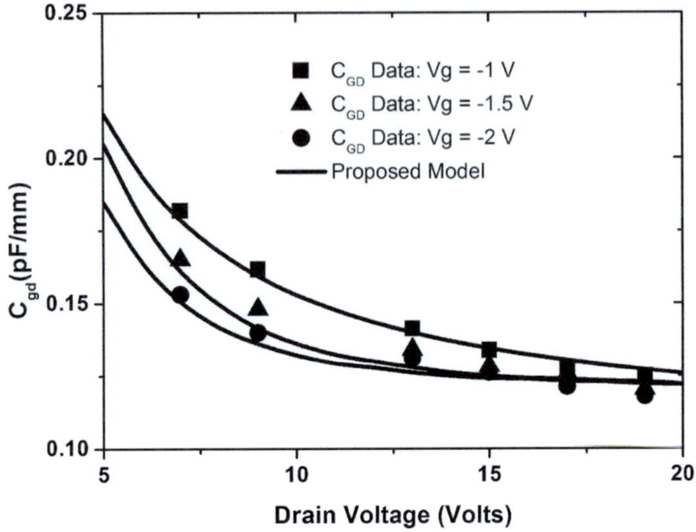

Fig. 7.2 Gate-to-drain capacitance versus V_{ds} of a GaN transistor without field-plates

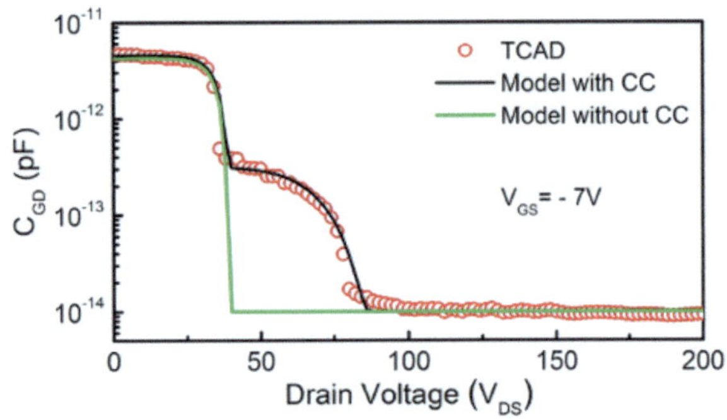

Fig. 7.3 Gate-to-drain capacitance versus V_{ds} of a GaN transistor with and without gate-connected field-plate. V_{ds} is varied from 0 to value beyond the magnitude of the cut-off voltage of the field-plate region

device. As V_{ds} is reduced and when V_{ds} becomes roughly equal to the magnitude of the cut-off voltage of the field-plate region 2-DEG starts to form in this region. As a result, C_{gd} starts to increase due to the 2-DEG under the gate-field region. This creates a hump like behavior in the C_{gd} versus V_{ds} characteristics.

The presence of source-connected field-plate can also create complexities in C_{gd} versus V_{ds} behavior. In Fig. 7.3 C_{gd} versus V_{ds} behavior of GaN device with gate- and source-connected field-plates is shown. Even for the device with source-connected field-plate a C_{gd} shows a hump-like behavior similar to device in which gate-connected field-plate was present. The hump occurs at the V_{ds} value equal to the magnitude of the source field-plate. This is due to the coupling of 2-DEG charge under the source field-plate to the gate, if the gate is physically close. This interesting capacitance bias dependence is captured in the ASM-HEMT model.

7.3.3 Drain–Source Capacitance

Drain–source capacitance is crucial for matching network design. The presence of field-plates in the device also affects drain–source capacitance. Typically drain–source capacitance C_{ds} decreases with increase in drain–source voltage as the depletion region width increases with it. The presence of field-plates causes charge densities to vary in the different regions of the device. This causes bias dependence in C_{ds}.

In Fig. 7.4 C_{ds} versus V_{ds} behavior of device with and without a source-connected field-plate is shown. For device without field-plate, C_{ds} monotonically decreases with an increase in V_{ds}. However, for device with source field-plate C_{ds} can show an increase as V_{ds} as V_{ds} is decreased to the cut-off voltage of the source field-plate region. As the source field-plate region becomes replete with 2-DEG due to the decrease in V_{ds} the drain–source capacitance C_{ds} increases.

The complex bias dependence of C_{ds} is known to cause non-linear AM/PM behavior (Fig. 7.5). Non-linear behavior is important for the RF power amplifier linearity. In Fig. 12.24 ASM-HEMT model simulations with and without source field-plate are shown. The presence of source field-plate causes non-linearity in the output capacitance. At large RF power the V_{ds} swing increases and can easily go beyond the cut-off voltage of the source field-plate region. This causes the C_{ds} non-linear behavior and gives rise to AM/PM increase.

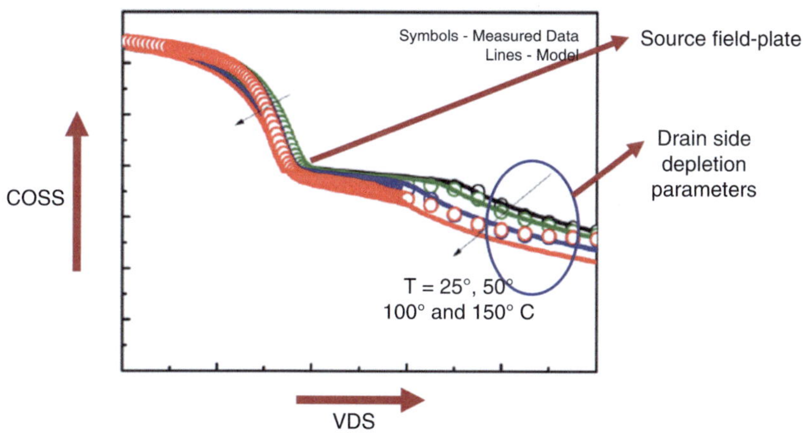

Fig. 7.4 Drain-to-source capacitance versus V_{ds} for device with and without source field-plate

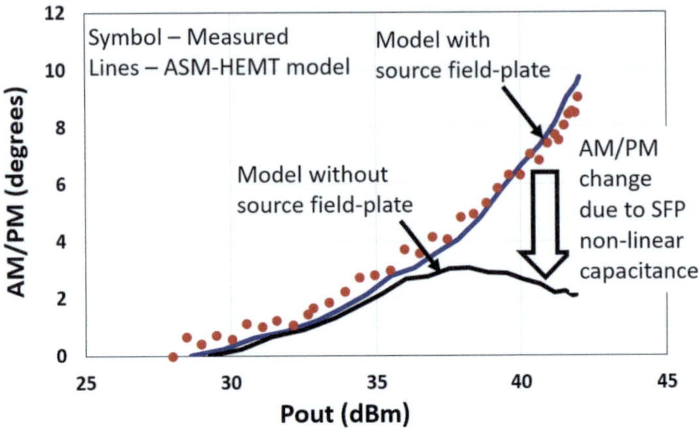

Fig. 7.5 Change in AM/PM non-linearity due to the presence of source field-plate. This non-linearity is accurately modeled with ASM-HEMT model

7.4 Impact of High Field Region on Capacitance

The electric field distribution in GaN HEMTs can be such that the horizontal component of the electric field peaks at the drain-side edge of the gate electrode. Due to the electric field, carriers in this region may reach their saturation velocity. The lateral extension, or length, of this high field region is directly affected by the value of V_{ds}. Such phenomena are typically referred to as channel-length modulation (CLM). The effects of CLM are predominantly seen in the Silicon (Si)-based material system, which can be attributed to Si being a more mature material system, thus having more aggressive scaling of the gate-length. The impact of CLM directly affects the intrinsic capacitances of the device. Such effects of CLM on the intrinsic capacitances have been widely researched in short-channel MOSFETs, where several modeling approaches have been reported to address this phenomenon [62–64].

GaN HEMTs have relatively long-channel lengths as compared to short-channel MOSFETs. Nevertheless, the large operating drain voltages are seen to cause high electric field regions where carrier velocity can reach saturation levels and in-turn, lead to CLM effects to occur. The electric field distribution near the drain-side edge of the gate electrode was analytically studied in [65].

A TCAD simulation of the device shown in Fig. 7.6 is performed to observe the electric field in the device. The electric field profile is shown in Fig. 7.7. From Fig. 7.7 it is clear that the charge which is present at the drain-side edge of the gate electrode sees a strong electric field which causes the mobile carriers to traverse through the channel at a much higher velocity than the carriers at the source-side edge of the gate electrode. This high electric field region is prone to increase to a level where the carrier velocity saturates.

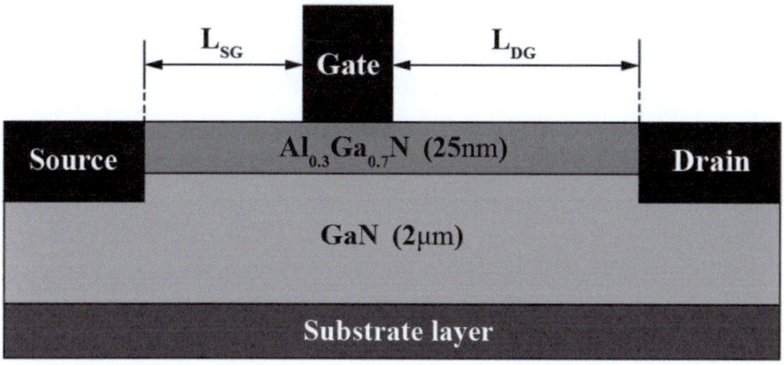

Fig. 7.6 GaN device simulated in TCAD to understand the electric field profile in the device

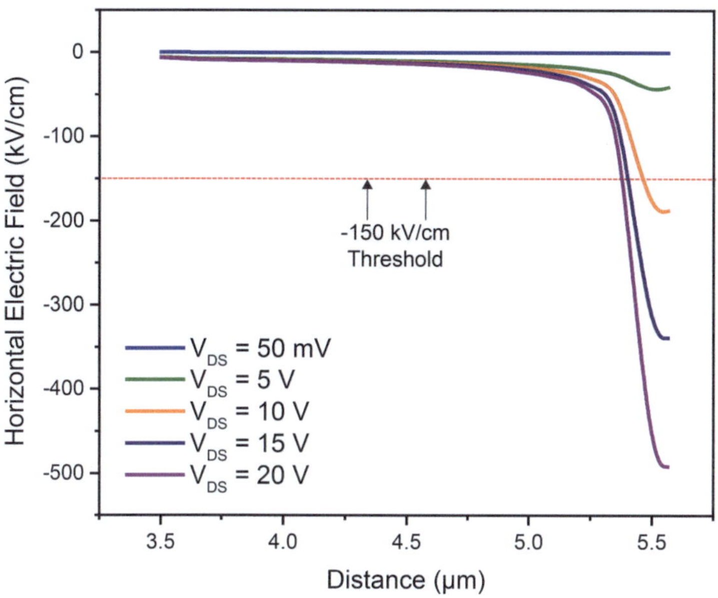

Fig. 7.7 Variation of electric field in the device moving from source to drain horizontally

It is made evident in Fig. 7.7 that the magnitude of the electric field is seen to be increasing with respect to increments of V_{ds}. Furthermore, for small values of V_{ds} (50 mV–5 V) the electric field ranges from being effectively 0 to -50 kV/cm, approximately. Notably, for values of V_{ds} greater than or equal to 10 V the threshold for the electric field value of -150 kV/cm has been surpassed, thus, it can be said that the velocity of the mobile carriers has reached saturation. The saturation of carrier velocity has a direct impact on the intrinsic capacitance behavior of the device. This is modeled in the following manner.

The total charge under the gate is calculated by integrating the two-dimensional electron gas (2-DEG) density from gate-edge toward source to the gate-edge toward drain. Khandelwal et al. [66] gives the following total gate charge equation for the ASM-GaN-HEMT model, which is reproduced here for ease of reference and is as follows:

$$Q_g = \frac{C_g L W}{\left(V_{go} - \psi_m + V_{th}\right)} \Big[V_{go}^2 + (1/3)\left(\psi_d^2 + \psi_s^2 + \psi_d \psi_s\right) - V_{go}\left(\psi_d + \psi_s - V_{th}\right) - V_{th}\psi_m \Big]. \tag{7.7}$$

Equation (12.2) is derived by neglecting the effects of charge carrier velocity saturation. However, as GaN-based technology improves, such effects are seen to be of significant importance for more aggressively scaled devices. Hence, an improved gate charge calculation is needed. It is shown in the literature [67] that the carrier saturation velocity for HEMTs can be written as

$$v_{sat} = \mu_{2DEG} E_{sat}. \tag{7.8}$$

Here v_{sat} is the saturation velocity, μ_{2DEG} is the carrier mobility in the 2-DEG, and E_{sat} is the electric field value at which the carrier velocity saturates. To model the influence of high field region, the gate charge is split into two different regions: Region I (Q_{g1}) and Region II (Q_{g2}), as depicted. Q_{g1} defines the region of the gate charge which is unaffected by the effects of velocity saturation and CLM. Q_{g2} defines the charge which is impacted by the effects of velocity saturation and CLM. As the bounds of the Q_{g1} region are beyond the confines of the CLM region, the charge calculation for the Q_{g1} region remains as given in (12.2) with the exception that L should be reduced by $L - L_{CLM}$ in order to account for the CLM region. The model for Q_{g1} was presented in the chapter on core model. For Q_{g2} region we examine the 1-D (horizontal) Poisson approximation for the Q_{g2} region. This gives,

$$\frac{\partial^2 V}{\partial y^2} = \frac{1}{\varepsilon} \frac{I_{ds}}{\alpha v_{sat}}, \tag{7.9}$$

where ε is the permittivity of AlGaN, I_{ds} is the drain-to-source current in the device, $\alpha = W \times \delta_{\mathrm{th}}$ with W as the gate-width and δ_{th} as the 2-DEG thickness. Using the current continuity equation, (7.9) can be easily integrated twice resulting in

$$\frac{\partial V}{\partial y} = \left(\frac{1}{\varepsilon} \frac{I_{ds}}{\alpha v_{sat}}\right) y + c_1, \tag{7.10}$$

$$V(y) = \left(\frac{1}{\varepsilon} \frac{I_{ds}}{\alpha v_{sat}}\right) y^2 + c_1 y + c_2, \tag{7.11}$$

where c_1 and c_2 are integration constants (7.11) takes the form of a standard quadratic equation. However, it is necessary to place certain boundary conditions in order to quantify the CLM region for Q_{g2}. Such boundary conditions, which have been depicted in Fig. 7.6, are as follows:

$$V\,(y = 0) = V_{di}, \tag{7.12}$$

$$\left.\frac{\partial V}{\partial y}\right|_{y=-L_{vsat}} = E_{sat}, \tag{7.13}$$

$$V\,(y = -L_{vsat}) = V_{dsat}, \tag{7.14}$$

where V_{di} is the intrinsic drain voltage, i.e., the voltage obtained after subtracting the voltage drop in the drain-side access region from the externally applied V_{ds}, L_{vsat} is the length of the CLM region, E_{sat} is the electric field value at which the carrier velocity saturates, and V_{dsat} is the drain-side saturation voltage. A boundary condition at V_{di}, or (7.12), is used in order to confine the right-side extremity of the Q_{g2} region. Similarly, the boundary condition of (7.14) determines the left-side extremity of the Q_{g2} region, which also captures the bias dependency of the proposed model. Note that (7.14) is an approximation as V_{dsat} is assumed to be independent of L_{vsat}. This approximation simplifies the model for compact modeling purposes. Equation (7.13) is used to quantify the electric field at which carrier velocity saturates. Incorporating (7.12)–(7.14) into (7.11) gives

$$\left[2\left(\frac{1}{\varepsilon}\frac{I_{ds}}{\alpha v_{sat}}\right)(-L_{vsat})\right]^2 - E_{sat}L_{vsat} = (V_{dsat} - V_{di}) \tag{7.15}$$

Equation (7.15) can be solved by means of the quadratic formula, allowing us to quantify L_{vsat} as follows:

$$L_{vsat} = \frac{-E_{sat} + \sqrt{{E_{sat}}^2 - 8\left(\frac{1}{\varepsilon}\frac{I_{ds}}{\alpha v_{sat}}\right)(V_{dsat} - V_{di})}}{4\left(\frac{1}{\varepsilon}\frac{I_{ds}}{\alpha v_{sat}}\right)}. \tag{7.16}$$

The above formulations are comparable to that of the collector epitaxial layer for a given bipolar transistor after base push-out occurs at high current density. This is commonly known as the Kirk effect. Traditional methods [68–70] of modeling this region in bipolar devices have been well established for a number of decades. By considering (7.14) to be equal to zero, (7.16) is seen to reduce to the depletion width of a single-sided PN junction. Therefore, the proposed methodology is inherently similar to that of the PN junction depletion region width calculation.

$$L_{vsat} = \frac{N_\varphi}{max\,(x, xi, \eta_i)}, \tag{7.17}$$

where N_φ is the numerator of (7.16), x is the denominator of (7.16), x_i = IMIN (minimum drain–source current parameter in ASM-HEMT model), η_i is a smoothing parameter. By incorporating the formulation of (7.17) into the terminal charge equations, the gate charge value for the Q_{g2} region can be ascertained, which is given as

$$Q_{g2} = \left(\frac{1}{\varepsilon}\frac{I_{ds}}{\alpha v_{sat}}\right) L_{vsat}. \tag{7.18}$$

Finally, we can calculate the total gate charge ($Q_{g\text{total}}$) with the following:

$$Q_{g\text{total}} = Q_{g1} + Q_{g2}, \tag{7.19}$$

where the updated calculation of Q_{g1} is given as

$$Q_{g1} = \frac{C_g W(L - L_{vsat})}{\left(V_{go} - \psi_m + V_{th}\right)} \left[V_{go}^2 + (1/3)\left(\psi_d^2 + \psi_s^2 + \psi_d\psi_s\right)\right. \tag{7.20}$$
$$\left. - V_{go}\left(\psi_d + \psi_s - V_{th}\right) - V_{th}\psi_m\right].$$

For the sample device in Fig. 7.6 the length of the CLM region with increasing V_{ds} is calculated in TCAD. The length of the CLM region in the TCAD simulations is calculated by applying the boundary conditions set forth by (7.12)–(7.14), the length of the CLM region can be manually calculated from the TCAD results. These values were then compared to the calculated values which were ascertained by using (7.17). The results are depicted in Fig. 7.8. The strong agreement between the simulations and the calculations verifies the proposed modeling formulations.

The impact of the high field region on the device capacitance can be seen from Figs. 7.9 and 7.10. Several observations can be made from Figs. 7.9 and 7.10. It can be seen that for linear conditions of V_{ds} (0–50 mV), there is no presence of the high electric field around the drain-side edge of the gate electrode. Thus, there is no impact to the C-V behavior. As V_{ds} increases to a value which is large enough that the carrier velocity reaches saturation, the impact of the CLM region begins to come into effect. It is here that the importance of a compact model which can capture this behavior is made evident. It is seen in Fig. 7.9 that V_{ds} has increased to a value such that the carrier velocity has reached saturation, and the effects of CLM are present, thereby impacting the C-V behavior. It is to be noted that by analyzing the results in Fig. 7.9 that the relative difference between the two models is seen to be slightly decreasing for an increase in V_{ds}. This can be attributed to the square-law behavior of the CLM region with respect to V_{ds}. In Fig. 7.10 it is shown that C_{gd} slightly decreases as compared to the previous model which does not include the Q_{g2} component in the total gate charge. This can be attributed to C_{gd} being defined as the rate of decrease of Q_g with respect to V_{ds} and the new component Q_{g2} increases with respect to V_{ds}, effectively reducing the rate of decrease of total Q_g, in-turn

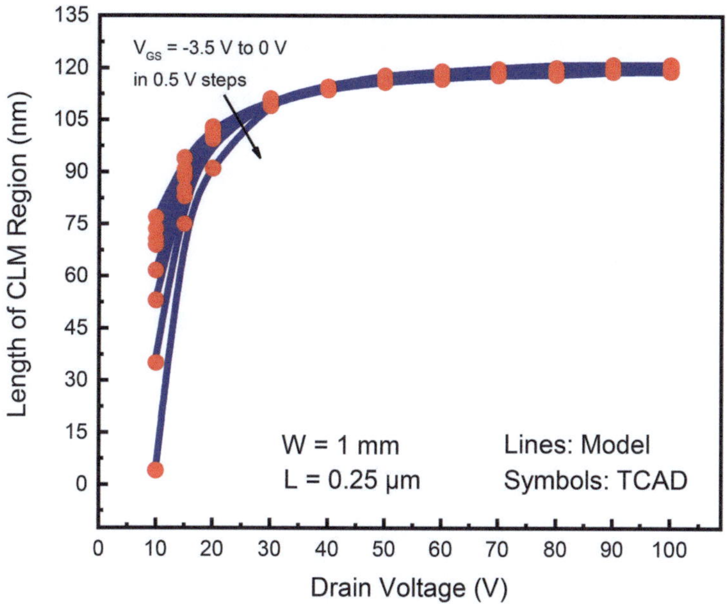

Fig. 7.8 Agreement between the developed modeling formulations and TCAD for the CLM for different drain and gate voltages

causing a decrease in C_{gd}. A similar effect on C_{gd} due to high field regions has been seen in silicon MOSFETs [63].

7.5 Impact of Charge Trapping on Capacitance

Charge trapping in buffer, interface, and barrier regions of the device is an important phenomenon in GaN HEMTs. The charge trapping causes change in device terminal characteristics as often measured by differences in pulsed and continuous-wave characteristics. The effect of charge trapping on the I-V characteristics is popularly known as drain-lag and gate-lag in the radio-frequency applications. This is known as dynamic on-resistance in power electronics applications.

Charge trapping occurs due to defects at the interface or in the semiconductor layers present in the GaN device. The defects can create trap centers within the band-gap. Trap centers can capture and release charges during the device operation. Trapped charges change the device electrostatics leading to change in both current and capacitance behavior. Change in drain current with trapping effects has been intensely researched, and several ways of modeling these have been developed. However, the effect of trapping on capacitance behavior of the device is also important. Moreover as the change in current and capacitance due to trapping occurs due to the same phenomenon of trapping, the modeling needs to be self-consistent.

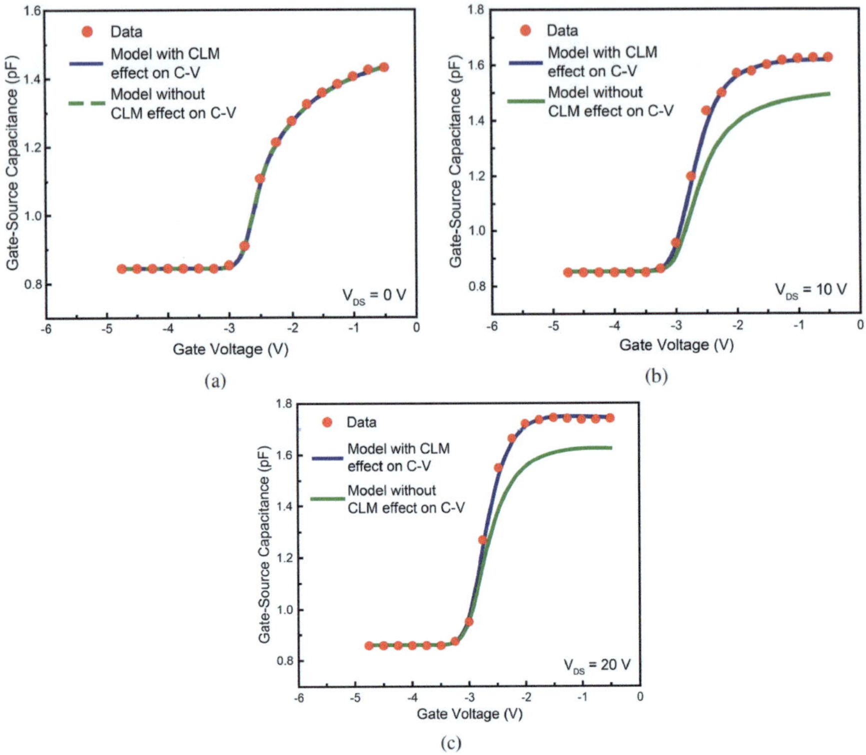

Fig. 7.9 Gate–source capacitance versus gate voltage comparison between a model which captures the impact of CLM effects against a model which neglects such effects. (**a**) $V_{ds} = 0$ V; (**b**) $V_{ds} = 10$ V; (**c**) $V_{ds} = 20$ V

This can be understood from Fig. 7.11. Figure 7.11 shows measured I-V and C-V characteristics of a GaN HEMT device. I-V from pulsed I-V measurements under three different drain quiescent conditions are shown, while C-V extracted from pulsed S-parameter measurements under the same drain quiescent voltage conditions are shown. A change in the cut-off voltage in I-V data is apparent from the top plot of Fig. 7.11. This change in the cut-off voltage is labeled as $\Delta V_{off,IV}$. Cut-off voltage can also be observed from the C_{gs} versus V_{gs} characteristics shown in the bottom plot of Fig. 7.11. The rise in C_{gs} occurs at the cut-off voltage, and a clear change in cut-off voltage under the different quiescent conditions is observed from Fig. 7.11. The change in the cut-off voltage in the C-V characteristics is labeled as $\Delta V_{off,CV}$. It can be observed that $\Delta V_{off,IV}$ is roughly equal to $\Delta V_{off,CV}$. By modeling the change in the cut-off voltage due to trapping effects. Self-consistent modeling of trapping effects is implemented in the ASM-HEMT model in similar manner as the I-V and C-V self-consistency. The general idea is that the key model parameters which get affected due to charge trapping are shared between current and charge formulations. As a result, the effect of charge trapping on model parameter

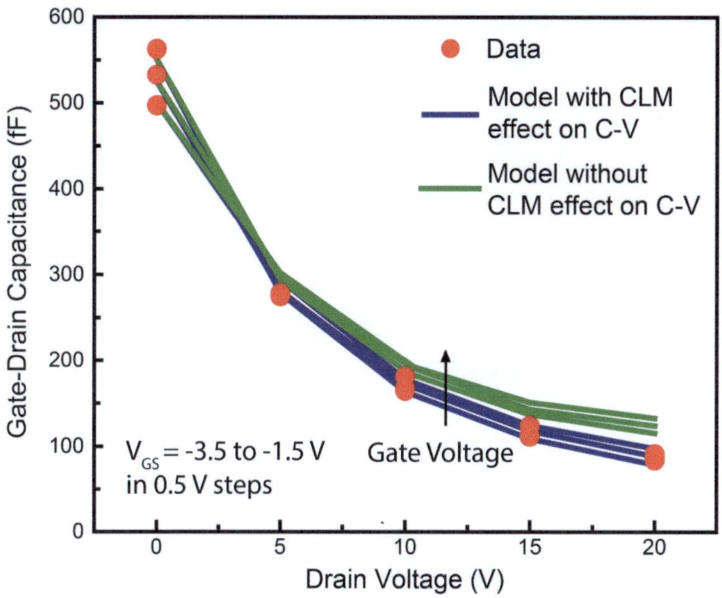

Fig. 7.10 C_{gd} versus V_{ds} under and the impact of modeling high field region charge on it

is modeled consistently in I-V and C-V behavior. The idea is shown in Fig. 7.12. This general idea is used in the internal trapping models implemented in the ASM-HEMT model, and this scheme can also be used to connect an external trapping model to the core ASM-HEMT model formulations.

7.6 Substrate Bias Effects and Substrate Capacitance

For power electronics GaN devices are fabricated on silicon substrates. In power converter switching applications, the dynamic switching of terminal voltage leads to effect of non-zero substrate voltage. To carefully model this scenario the effect of substrate voltage should be considered. A typical GaN device with back/substrate metallization is shown in Fig. 7.13.

The substrate voltage dependence of the 2-DEG charge in each region of the device needs to be modeled. The electric field lines from the substrate modify the 2-DEG charge formulation from Eqs. (7.21) to (7.22).

$$qn_s = C_g\left(V_g - V_{off}\right). \tag{7.21}$$

$$qn_s = C_g\left(V_g - V_{off}\right) + C_b\left(V_b - V_{off,b}\right). \tag{7.22}$$

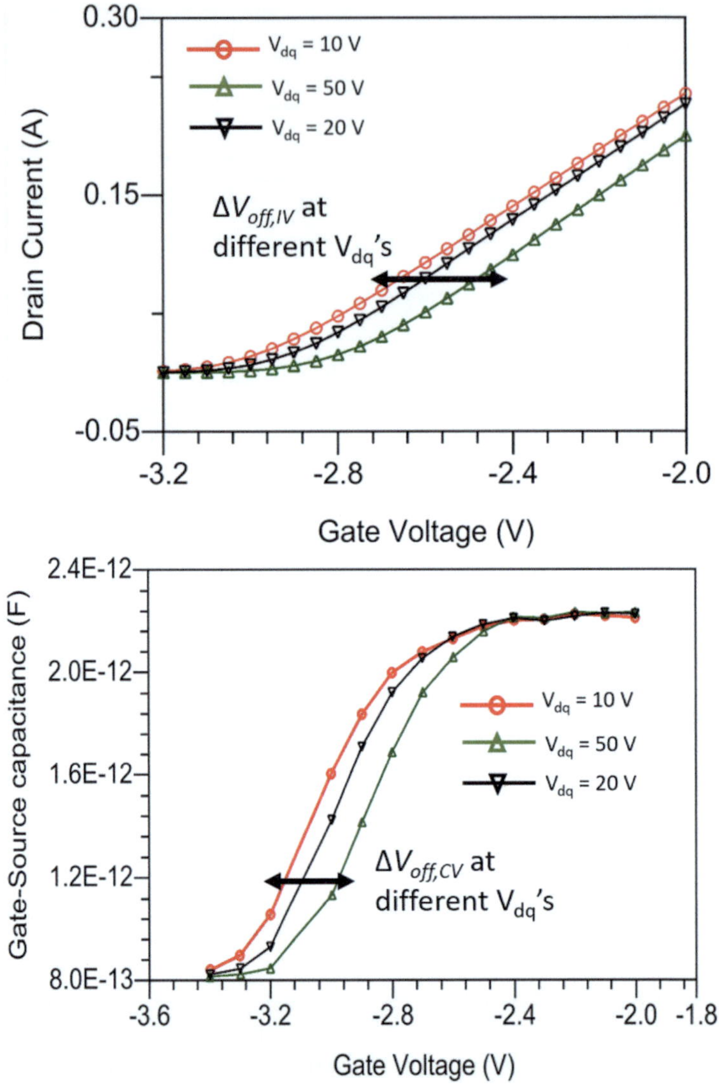

Fig. 7.11 C_{gd} versus V_{ds} under and the impact of modeling high field region charge on it

The above formulations can be re-written as,

$$qn_s = C_t \cdot \left(\frac{C_g}{C_t} V_g - \frac{C_g}{C_t} V_{off} \right) + C_t \left(\frac{C_b}{C_t} V_b - \frac{C_b}{C_t} V_{off,b} \right), \tag{7.23}$$

where $C_t = C_b + C_g$. This can be simplified to Eq. (7.24)

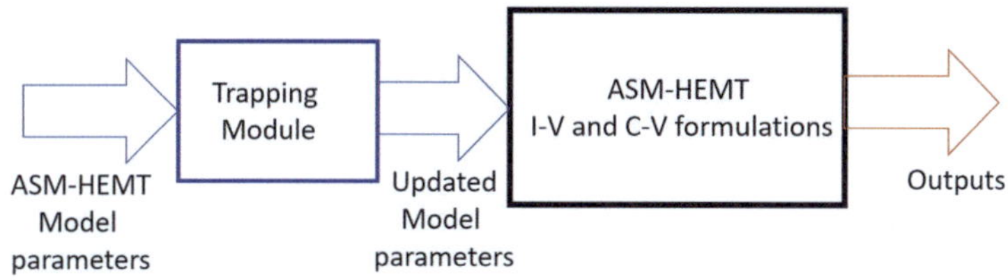

Fig. 7.12 Block diagram representation showing how trapping effects are accounted for in the ASM-HEMT model self-consistently for I-V and C-V model behavior

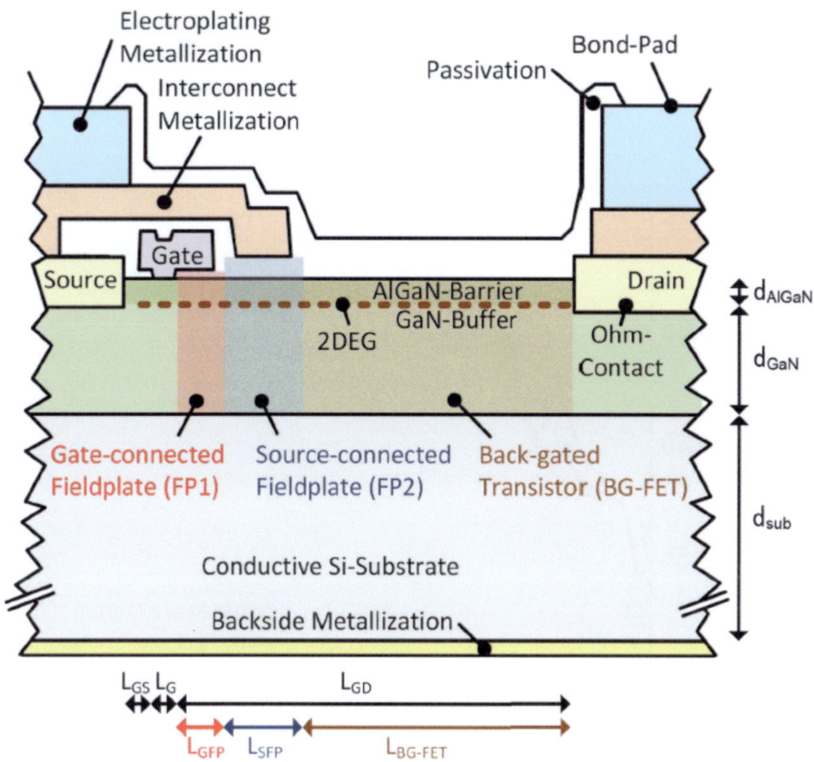

Fig. 7.13 GaN device with back metallization at the substrate. Field-plate and access regions of the device are labeled. All regions get affected by the substrate voltage

$$qn_s = C_t \cdot \left(\frac{C_g}{C_t} V_g - \frac{C_g}{C_t} V_{off} + \frac{C_b}{C_t} V_b - \frac{C_b}{C_t} V_{off,b} \right). \tag{7.24}$$

The formulation Eq. (7.24) accounts for the capacitive coupling due to the substrate voltage. 2-DEG under each region of the device gets affected due to the substrate coupling and can be accounted for with Eq. (7.24). This can be achieved by calculating effective gate voltage $V_{g,eff}$ which goes into the surface-potential core calculation. $V_{g,eff}$ can be calculated as,

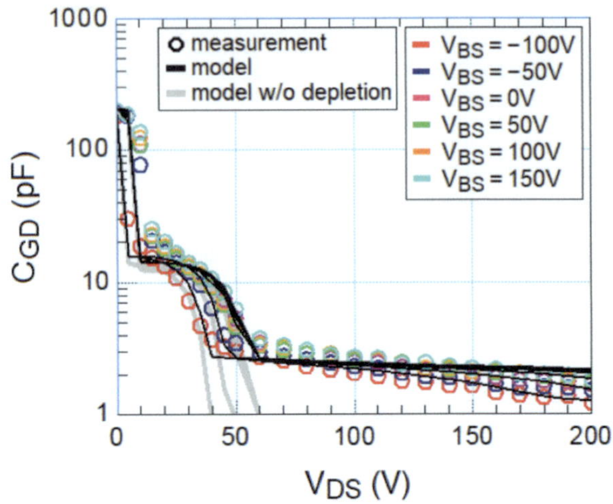

Fig. 7.14 Modeling the impact of substrate voltage on gate–drain capacitance

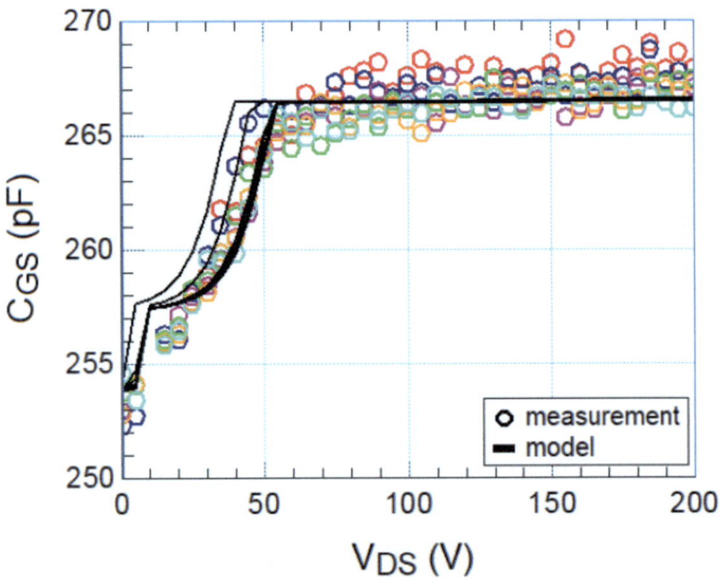

Fig. 7.15 Modeling the impact of substrate voltage on gate–source capacitance

$$V_{g,eff} = \frac{C_g}{C_t} \cdot V_g + \frac{C_b}{C_t} \cdot V_b. \tag{7.25}$$

$V_{g,eff}$ then is used in surface-potential calculations in each region of the GaN device. These updated formulations have been used to model the effect of a GaN device with two field-plates. Accurate model for capacitance has been achieved as shown in Figs. 7.14 and 7.15.

Chapter 8
Gate Current Model

8.1 Diode-Based Formulation

The modeling of gate current in ASM-HEMT is implemented by using two diodes: one is for gate-to-source current and another is for gate-to-drain current. The current expressions used in GATEMOD=1 are given below.

$$I_{gs} = W \cdot L \cdot NF \cdot \left[IGSDIO + \left(\frac{T_{dev}}{TNOM} - 1 \right) \cdot KTGS \right] \times \left[exp \left\{ \frac{V_{gs}}{NJGS \cdot K_B \cdot T_{dev}} \right\} - 1 \right] \tag{8.1}$$

$$I_{gd} = W \cdot L \cdot NF \cdot \left[IGDDIO + \left(\frac{T_{dev}}{TNOM} - 1 \right) \cdot KTGD \right] \times \left[exp \left\{ \frac{V_{gd}}{NJGD \cdot K_B \cdot T_{dev}} \right\} - 1 \right], \tag{8.2}$$

where *IGSDIO* and *IGDDIO* represent saturation currents of gate–source and gate–drain junction diodes, respectively. *NJGS* and *NJGD* are the diode current ideality factors, and parameters *KTGS* and *KTGD* are used to capture the temperature dependence of gate saturation current.

GATEMOD=1 is a simple diode-like reverse current model. GaN HEMTs can have large reverse leakage due to Poole–Frenkel effect. The following formulations

S. Khandelwal, *Advanced SPICE Model for GaN HEMTs (ASM-HEMT)*,
https://doi.org/10.1007/978-3-030-77730-2_8

are implemented in the model for Poole–Frenkel reverse current. These get invoked when GATEMOD=2.

$$I_{gs,dio} = W \cdot L \cdot NF \cdot \left[IGSDIO + \left(\frac{T_{dev}}{TNOM} - 1 \right) \cdot KTGS \right] \times \left[exp\left\{ \frac{V_{gs}}{NJGS \cdot K_B \cdot T_{dev}} \right\} - 1 \right]. \tag{8.3}$$

8.2 Model Based on Frenkel–Poole Mechanism

The following additional formulations are added for Poole–Frenkel leakage.

$$E_{field} = Vgs/TBAR \tag{8.4}$$

$$RIGSDIO_T = RIGSDIO + exp(T/TNOM - 1) * RKTGS \tag{8.5}$$

$$I_{gs} = I_{gs,dio} \cdot \left(1 + RIGSDIO_T * E_{field} \cdot exp\left(\frac{\sqrt{Vgs} + EBREAKS}{RNJGS * K_B * T} \right) \right). \tag{8.6}$$

The above equations are for gate–source part of the gate current. Similar equations exist in the model for the gate–drain part of the total gate current.

8.3 Gate Current with Consistency with Drain Current

The reverse-bias gate current is mainly governed by the Poole–Frenkel (PF) emission and Fowler–Nordheim tunneling [71] at medium to high reverse bias, whereas thermionic emission (TE) plays a vital role in the forward-bias region [71–75]. These current components together attribute the gate current for a wide bias range. In this section, we shall develop an accurate model for these three components of the gate current within the framework of the ASM-GaN compact model.

8.3.1 Thermionic Emission Model

In a Schottky contact, thermionic emission (TE) is the dominant conduction mechanism in the forward-bias range. The current density–voltage characteristics of a Schottky contact (or diode) is given by Sze [76]

$$J_{TE} = J_{TE0}\ (\exp[\,V/(\eta\, V_{th})\,] - 1)\,, \tag{8.7}$$

where V is the applied voltage, η is the ideality factor, $V_{th} = K_B\, T$ is the thermal voltage, where K_B is the Boltzmann's constant and T is the temperature, and J_{TE0} is the reverse saturation current density and is given by

$$J_{TE0} = A^*\, T^2\, \exp(-q\, \phi_{TE}/V_{th})\,, \tag{8.8}$$

where ϕ_{TE} is the Schottky barrier height and A^* is the effective Richardson's constant, which is given by Sze [76]

$$A^* = 4\pi\, q\, m^*\, k_B^2/h^3\,, \tag{8.9}$$

where h is the Planck's constant, k_B is the Boltzmann's constant, and m^* is the electronic effective mass in AlGaN barrier.

In a GaN HEMT, the applied voltage varies along the channel, from the source end of the channel toward the drain end of the channel. To accurately model the TE current, a distributed network of Schottky diodes is needed. This can be achieved by modeling the intrinsic region of the GaN HEMT as a series of transistors, and associating with each transistor a Schottky diode, which is modeled by integrating (8.7) along the channel length and width of each transistor. In [75], this complex structure is simplified by integrating (8.7) along the channel length and width to obtain the TE current, I_{TE}. The gate-to-source and gate-to-drain TE current components, denoted respectively as I_{TES} and I_{TED}, are then implemented using the 60:40 partitioning scheme, i.e., $I_{TES} = 0.6\ I_{TE}$ and $I_{TED} = 0.4\ I_{TE}$. An alternative, but simpler, model is to consider two diodes: one for the gate-to-source TE current and another for the gate-to-drain TE current. The two current components, denoted respectively as I_{TES} and I_{TED}, are obtained by integrating J_{TE} over the gate length and width, setting the applied gate voltage to, respectively, the intrinsic gate-to-source voltage, $V_{G_i S_i}$, and the intrinsic gate-to-drain voltage, $V_{G_i D_i}$. The resulting TE current components are

$$I_{TE(S,D)} = \frac{1}{2}\ NF\ W\ L\ J_{TE0}\left[\exp\left(\frac{V_{G_i(S_i,D_i)}}{\eta\, V_{th}}\right) - 1\right], \tag{8.10}$$

where NF is the number of fingers. The pre-factor 0.5 in (8.10) accounts for the fact that ideally at $V_{DS} = 0$ V, $I_{TES} = I_{TED} = I_{TE}/2$.

To account for the simplifications encountered in the above calculations, we introduce the following enhancements in the model. First, the gate-to-source voltage $V_{G_i S_i}$ and the gate-to-drain voltage $V_{G_i D_i}$ in (8.10) are replaced with the following effective voltages:

$$V_{GS,eff} = \beta_S\ V_{G_i S_i} + (1 - \beta_S)\ V_{G_i D_i} \tag{8.11}$$

$$= V_{G_i D_i} + \beta_S\ V_{D_i S_i}\,, \tag{8.12}$$

$$V_{GD,eff} = \beta_D \, V_{G_i S_i} + (1 - \beta_D) \, V_{G_i D_i} \tag{8.13}$$

$$= V_{G_i D_i} + \beta_D \, V_{D_i S_i} , \tag{8.14}$$

where the model parameter $0 \leq \beta_S \leq 1$, hence, $V_{G_i D_i} \leq V_{GS,eff} \leq V_{G_i S_i}$, and the model parameter $0 \leq \beta_D \leq 1$, hence, $V_{G_i D_i} \leq V_{GD,eff} \leq V_{G_i S_i}$. Note that introducing these empirical parameters results in $V_{G_i D_i}$ dependence of the gate-to-source leakage current and $V_{G_i S_i}$ dependence of the gate-to-drain leakage current. Second, the Schottky barrier height ϕ_{TE} at the drain side of the channel is slightly altered with respect to the one at the source side of the channel as

$$\phi_{TES} = \phi_{TE} , \tag{8.15}$$

$$\phi_{TED} = \phi_{TE} + \Delta\phi , \tag{8.16}$$

where ϕ_{TES} and ϕ_{TED} are the source- and drain-side Schottky barrier heights, and $\Delta\phi$ is a model parameter. Note that this modification of the Schottky barrier height also accounts for possible non-idealities in the structure of the GaN HEMT that can affect the Schottky barrier height along the channel.

Our analysis of the measured characteristics of the devices under test shows, in agreement with the results reported in the literature [74, 77], that accurate modeling of these characteristics requires inclusion of temperature dependence for the Schottky barrier height ϕ_{TE} and the ideality-factor parameter η as

$$\phi_{TE,T} = \phi_{TE} + K_{\phi_{TE}} \, (T/T_n - 1) , \tag{8.17}$$

$$\eta_T = \eta + K_\eta \, (T/T_n - 1) , \tag{8.18}$$

where T_n is the nominal temperature and $K_{\phi_{TE}}$ and K_η are the model parameters.

8.3.2 *Poole–Frenkel Model*

The thermionic emission current over the Schottky barrier is too small to explain the observed excess gate leakage current in AlGaN/GaN HEMTs. This excess leakage current is, rather, attributed either partially or completely to the leakage path in III-nitride semiconductors resulting from the presence of threading dislocations in the AlGaN barrier. The conductive nature and the carrier-capture radius of each individual dislocation depend on factors, such as types and distribution of dislocations and III/V ratios during epitaxial growth [72]; therefore, it is difficult to incorporate the impact of threading dislocations into a gate leakage model. The combined effect of dislocations is, rather, modeled as a continuous band of conductive states (or a continuum of states). The gate leakage in the reverse-bias condition is then modeled as the flow of the electrons that are emitted from trap states onto this continuum of states, from the gate contact toward the channel

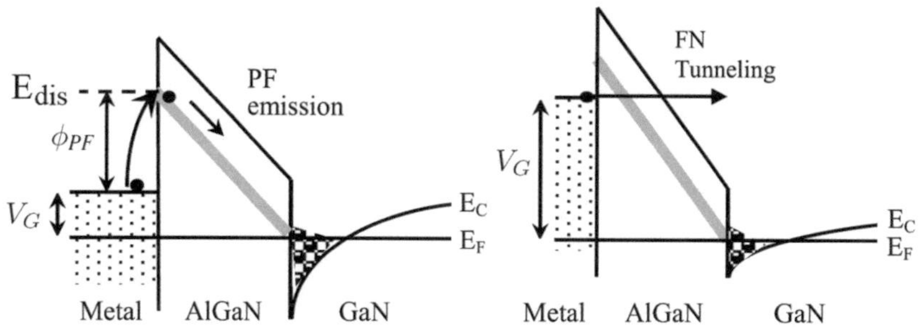

Fig. 8.1 Conduction band edge diagram of AlGaN/GaN HEMT for medium reverse gate voltage [71]. The left-hand side energy band diagram describes the physical mechanism of the PF emission at a medium applied gate bias. The trap state is assumed to be very close to the metal Fermi level, and the energy level associated with the continuum of states, denoted as E_{dis} and marked in gray, is at a height ϕ_{PF} from the metal Fermi level, which varies with the electric field. The right-hand side figure describes the Fowler–Nordheim tunneling process

through the AlGaN barrier [72], as shown in Fig. 8.1. The flow of electrons is caused by the vertical electric field $E_{\perp}$ in the AlGaN layer resulting from the application of the gate voltage. In addition to causing the flow of electrons, the electric field in the AlGaN layer also enhances the emission of the electrons from the trap states onto the continuum of states. This mechanism is called the Poole–Frenkel (PF) mechanism.

The relation between current density (J_{PF}) and electric field ($E_{\perp}$) in the PF conduction is given by Sze [76]

$$J_{PF} = E_{\perp}\, C_{PF}\, \exp\left(-q\, \phi_{PF}/V_{th}\right), \tag{8.19}$$

where C_{PF} is a model parameter and

$$\phi_{PF} = \phi_{PF0} - \sqrt{q\, E_{\perp}/\left(\pi\, \tilde{\epsilon}\right)}, \tag{8.20}$$

where ϕ_{PF0} is the zero-electric-field barrier height, and $\tilde{\epsilon}$ is the high frequency dielectric permittivity of the AlGaN layer and is given by

$$\tilde{\epsilon} = \kappa\, \epsilon, \tag{8.21}$$

where κ is a model parameter. As can be noticed, the potential barrier ϕ_{PF} decreases with increase in electric field.

The calculation of the PF current, I_{PF}, begins with calculating the electric $E_{\perp}$ as [73]

$$E_{\perp} = \left[q\, \sigma_p - C_g\, \left(V_{G0} - \psi\right)\right]/\epsilon, \tag{8.22}$$

where σ_p is the sum of the piezo-electric polarization charge in the barrier and the difference between spontaneous polarization charge in the barrier and the buffer, C_g

is the gate capacitance per unit area ($= T_{BAR}/\epsilon$, where T_{BAR} is the AlGaN thickness), and V_{G0} is the gate overdrive, which equals the difference between the applied gate voltage and the work-function difference (in volt) between the gate contact and the AlGaN layer, V_{OFF}:

$$V_{G0} = V_G - V_{OFF}. \tag{8.23}$$

This expression for electric field accounts for the presence of the strong polarization field within the AlGaN barrier and the screening effect of the two-dimensional electron gas (2DEG) as well as the ionized surface states.

The two PF current components, denoted respectively as I_{PFS} and I_{PFD}, are obtained, following the same modeling approach as of the TE current, by considering two PF current sources; the gate-to-source PF current source and the gate-to-drain PF current source are obtained by integrating J_{PE} over the gate length and width, setting the applied gate voltage to, respectively, the intrinsic gate-to-source voltage, $V_{G_iS_i}$, and the intrinsic gate-to-drain voltage, $V_{G_iD_i}$, and hence, yielding the following relations for the vertical electric fields at the source and drain ends of the channel, $E_{\perp,S}$ and $E_{\perp,D}$:

$$E_{\perp,(S,D)} = \left[q\,\sigma_p - C_g\,\left(V_{G_i(S_i0,D_i0)} - \psi_{(S_i,D_i)}\right)\right]/\epsilon\,, \tag{8.24}$$

where $V_{G_iS_i0} = V_{G_iS_i} - V_{OFF}$ and $V_{G_iD_i0} = V_{D_iS_i} - V_{OFF}$. The source-side and drain-side surface-potentials ψ_{S_i} and ψ_{D_i} are obtained directly from the ASM-GaN compact model. The surface-potential ψ_{D_i} is calculated using the same formulation as of ψ_{S_i}, with the $V_{G_iD_i}$ in place of $V_{G_iS_i}$, which is needed to calculate ψ_{S_i}. The two PF current components, I_{PFD} and I_{PFS}, are then given by

$$I_{PF(S,D)} = \frac{1}{2}\,NF\;W\,L\,E_{\perp,(S,D)}\,C_{PF}\,\exp\left(-\frac{\phi_{PF(S,D)}}{V_{th}}\right), \tag{8.25}$$

where

$$\phi_{PF(S,D)} = \phi_{PF0} - \sqrt{q\,E_{\perp,(S,D)}/\left(\pi\tilde{\epsilon}\right)}\,. \tag{8.26}$$

To account for the simplifications encountered in the above calculations, we introduce the following enhancements in the model. First, the gate-to-source voltage $V_{G_iS_i}$ and the gate-to-drain voltage $V_{G_iD_i}$ used in the calculation of the vertical electric fields $E_{\perp,(S,D)}$, given by (8.24), are replaced with the effective voltages given by (8.12) and (8.14), and the surface-potentials ψ_{S_i} and ψ_{D_i} are replaced with the effective surface-potentials given by

$$\psi_{S_i,eff} = \beta_S\,\psi_{S_i} + (1 - \beta_S)\,\psi_{D_i}\,, \tag{8.27}$$

$$\psi_{D_i,eff} = \beta_D\,\psi_{S_i} + (1 - \beta_D)\,\psi_{D_i}\,, \tag{8.28}$$

where the model parameters $0 \leq \beta_S \leq 1$ and $0 \leq \beta_D \leq 1$ are the same ones used in (8.12) and (8.14). Second, following the same enhancement in the Schottky barrier height implemented in the TE model, here also we modify the source and drain barrier heights for electron emission from the trap states to the continuum of states as

$$\phi_{PFS0} = \phi_{PF0}, \tag{8.29}$$

$$\phi_{PFD0} = \phi_{PF0} + \Delta\phi, \tag{8.30}$$

where $\Delta\phi$ is the same offset potential used in (8.16). Third, the work-function difference V_{OFF} used in (8.23) is slightly altered as

$$\widetilde{V}_{OFF} = V_{OFF} + \Delta V_{OFF}, \tag{8.31}$$

where ΔV_{OFF} is a model parameter.

Our analysis of the measured characteristics of the devices under test shows that accurate modeling of these characteristics requires inclusion of temperature dependence for the barrier height ϕ_{PF0} as

$$\phi_{PF0,T} = \phi_{PF0} + K_{\phi_{PF}}\,(T/T_n - 1), \tag{8.32}$$

where $K_{\phi_{PF}}$ is a model parameter.

Equation (8.24) yields non-zero vertical electric fields at zero biasing condition, i.e., at $V_{GS} = V_{DS} = 0$ V. Consequently, (8.25) yields non-zero PF current at zero biasing condition. Unless there exists a balancing conduction mechanism, the above formulation results in a non-physical situation at zero biasing condition. In [73, 75], a trap-assisted tunneling mechanism has been considered as the balancing mechanism, which balances out the PF current at zero biasing condition. One can alternatively attribute this non-physical behavior to the "phenomenological" modeling of the combined effect of dislocations in the AlGaN layer as a continuum of conductive states. Irrespective of the origin of this non-physical behavior, we shall consider the following re-calculation of the two components of the PF current to overcome this problem from the perspective of compact modeling:

$$I_{PF(S,D)} \rightarrow \xi_{(S,D)}\left(I_{PF(S,D)} - I_{PF(S0,D0)}\right), \tag{8.33}$$

where I_{PFS0} and I_{PFS0} are the gate-to-source and gate-to-drain PF currents calculated at zero biasing condition, and ξ_S and ξ_D are given by

$$\xi_{(S,D)} = \exp\left[\min\left(V_{G_i(S_i,D_i)}, 0\right)/V_{th}\right] - 1, \tag{8.34}$$

where $\min\left(V_{G_i(S_i,D_i)}, 0\right)$ refers to taking the minimum of $V_{G_i(S_i,D_i)}$ and 0 V. Subtraction of $I_{PF(S0,D0)}$ ensures zero PF current at zero biasing condition, at all temperatures, and multiplication by $\xi_{(S,D)}$ ensures zero first and higher derivatives of the PF with respect to V_{GS} and V_{GD} at zero biasing condition, at all temperatures.

8.3.3 Fowler–Nordheim Model

As the gate voltage becomes more negative, the electric field across the AlGaN barrier increases. At a critical electric field, E_c, when the barrier width at the metal Fermi level becomes $< d_c$, the electrons at the metal Fermi level tunnel through the AlGaN barrier, as shown in Fig. 8.1, yielding an additional component for the gate current, namely, the Fowler–Nordheim (FN) current. The J_{FN}-E dependence of FN tunneling is given by Sze [76]

$$J_{FN} = q\ \mu\ n_{FN}\ E_{\perp}\,, \tag{8.35}$$

where q is the electron charge, μ is the electron drift mobility in the AlGaN layer, $E_{\perp}$ is the vertical electric field in the AlGaN layer, and n_{PF} is the electron density resulting from the FN mechanism and is given by

$$n_{FN} = D\ \Delta E_{\perp}\ \exp\left(-\ B/\Delta E_{\perp}\right), \tag{8.36}$$

where D is a constant,

$$\Delta E_{\perp} = E_{\perp} - E_c\,, \tag{8.37}$$

$$B = 8\pi\ \sqrt{2\,m^*\ (\phi_{FN})^3}/\left(3\,q\,h\right)\,, \tag{8.38}$$

where m^* is the electronic effective mass in the AlGaN barrier, h is the Planck's constant, and ϕ_{FN} is the effective barrier height and can be estimated using the relation

$$E_c \approx \phi_{FN}/d_c\,, \tag{8.39}$$

where the critical barrier width d_c is the barrier width at which the FN process initiates, and is a few nano meters. Combining (8.35) and (8.36),

$$J_{FN} = E_{\perp}\,C_{FN}\ \Delta E_{\perp}\ \exp\left(-B/\Delta E_{\perp}\right), \tag{8.40}$$

where $C_{FN} = q\,\mu\,D$ is a model parameter.

Following the same modeling approach as of the TE and PF currents, we consider two FN current sources: the gate-to-source FN current source and the gate-to-drain FN current source. The two current components, denoted respectively as I_{FNS} and I_{FND}, are obtained by integrating J_{FN} over the gate length and width, setting the applied gate voltage to, respectively, the intrinsic gate-to-source voltage, $V_{G_iS_i}$, and the intrinsic gate-to-drain voltage, $V_{G_iD_i}$, and yielding the following relations for I_{FNS} and I_{FND}:

$$I_{FN(S,D)} = \frac{1}{2} NF\ W\ L\ E_{\perp,(S,D)} \times C_{FN}\ \Delta E_{\perp,(S,D)} \exp\left(\frac{-B}{\Delta E_{\perp,(S,D)}}\right). \tag{8.41}$$

The vertical electric fields $E_{\perp,(S,D)}$ are calculated in the same way as for the calculation of the two components of the PF current. Following the same enhancement implemented in the potential barrier heights used in the TE and PF models, here also we modify the effective barrier height ϕ_{FN} as

$$\phi_{FNS} = \phi_{FN}, \tag{8.42}$$

$$\phi_{FND} = \phi_{FN} + \Delta\phi. \tag{8.43}$$

Our analysis of the measured characteristics of the devices under test shows that accurate modeling of these characteristics requires inclusion of temperature dependence for the effective barrier height ϕ_{FN} as, in agreement with the results reported in the literature [71]:

$$\phi_{FN,T} = \phi_{FN} + K_{\phi_{FN}}\,(T/T_n - 1)\,, \tag{8.44}$$

where $K_{\phi_{FN}}$ is a model parameter. Note that with this relation, the critical electric field E_c, given by (8.39), also becomes temperature dependent.

In order to ensure zero FN current and zero first and higher derivatives of the FN current with respect to V_{GS} and V_{GD} at zero biasing condition, at all temperatures, we re-calculate the two FN components, in a similar manner to as we did for the two components of the PF current, as

$$I_{FN(S,D)} \rightarrow \xi_{(S,D)}\left(I_{FN(S,D)} - I_{FN(S0,D0)}\right)\,, \tag{8.45}$$

where I_{FNS0} and I_{FNS0} are the gate-to-source and gate-to-drain PF currents calculated at zero biasing condition, and ξ_S and ξ_D are given by (8.34).

The extracted parameters for the thermionic emission current are tabulated in Table 8.1. Using the theoretical value for the effective Richardson's constant (= 26.4 A cm^{-2} K^{-2}, corresponding to an electronic effective mass of 0.22 m_0, where m_0 is the electron mass), the extracted value for the Schottky barrier height at T = 25°C, ϕ_{TE}, is 0.94 eV, which is close to the values reported in the literature [72, 75, 78, 79]. As shown in Fig. 8.2, $\phi_{TE,T}$, given by (8.17), increases with increase in temperature, from 0.94 eV at T = 25°C to 1.19 eV at T = 200°C, hence, shows a 27.7% increase in its value over this temperature range. On the contrary, the ideality factor η_T, given by (8.18), shows, as also shown in Fig. 8.2, a 20.9% decrease in its value over the same temperature range. These trends have already been reported in

Table 8.1 Thermionic emission (TE) model parameters

	Description	Value
A^*	Effective Richardson's constant ($Acm^{-2}K^{-2}$)	26.4
ϕ_{TE}	Schottky barrier height at 25°C (eV)	0.94
$K_{\phi_{TE}}$	Temperature parameter for ϕ_{TE} (eV)	0.43
$\Delta\phi$	Potential offset relating the source side ϕ_{TE}, ϕ_{TES}, to the drain side ϕ_{TE}, ϕ_{TED} (eV)	46e−3
η	Ideality factor for the TE current	1.41
K_η	Temperature parameter associated with η	−0.5
β_S	Weighting factor used in the calculation of the effective gate-to-source potential	0.7
β_D	Weighting factor used in the calculation of the effective gate-to-drain potential	0

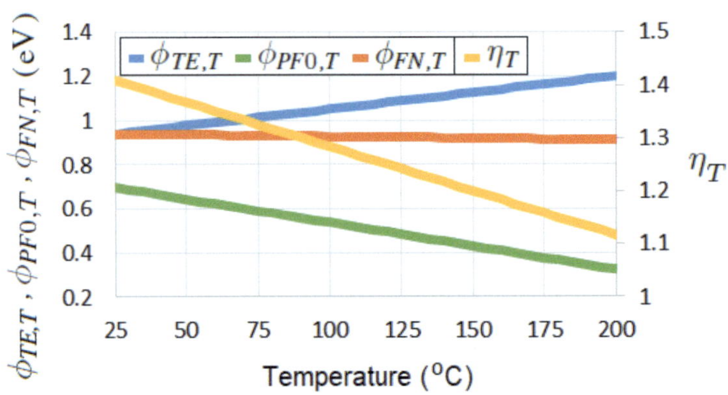

Fig. 8.2 The variations, with temperature, of the Schottky barrier height $\phi_{TE,T}$ and the ideality factor η_T used in the thermionic emission model, the barrier height for electron emission from the trap state to the continuum of states, $\phi_{PF0,T}$, used in the Poole–Frenkel model, and the effective barrier height, $\phi_{FN,T}$, used in the Fowler–Nordheim model

the literature [74, 77]. The excellent fit of the proposed model to the measured gate current under forward-bias condition is shown in Fig. 8.3.

As can be noted from Table 8.1, the extracted values for the parameters $0 \leq \beta_S, \beta_D \leq 1$ in (8.12) and (8.14), used in the model for the thermionic emission current, are 0.7 and 0, respectively, which ideally are 1 and 0, respectively. In order to show the effect of the parameter β_S, which deviates from the ideal value, we simulated the model for the device with gate length $L = 5\,\mu m$ at $V_{DS} = 0.5$ V, at $T = 25°C$, setting this parameter to 0 and 1. The results are shown in Fig. 8.3.

The extracted parameters used in the calculation of the vertical electric field in the AlGaN barrier are tabulated in Table 8.2. The extracted value for the polarization charge σ_p is $2.33 \times 10^{13}\,cm^{-2}$, which is in the same range of values reported in the literature [72, 74, 75]. From (8.24), σ_p enters in the calculation of the vertical electric fields at the source and drain sides of the channel, $E_{\perp,(S,D)}$. Figure 8.4

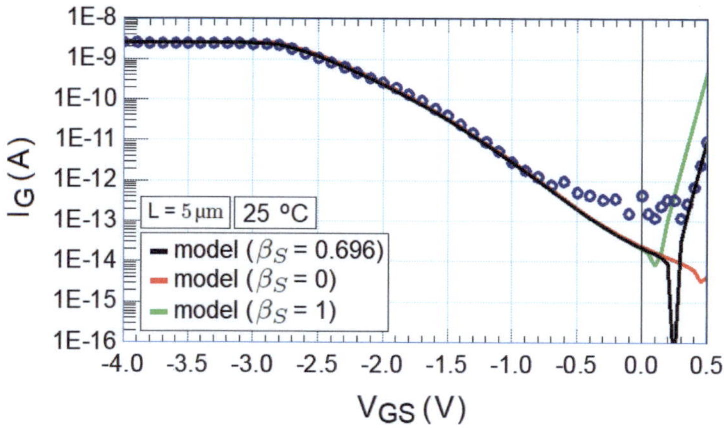

Fig. 8.3 Measured (symbol) and simulated (line) gate current (I_G)-gate–source voltage (V_{GS}) characteristics in semi-logarithmic scale for the device with gate length $L = 5\,\mu$m at the drain-to-source voltage $V_{DS} = 0.5$ V, at ambient temperature $T = 25°$C. Simulation results are shown for when the parameter β_S in (8.12) is set to 0 and 1

Table 8.2 Parameters used in the calculation of the vertical electric field in the AlGaN barrier

	Description	Value
σ_p	Polarization charge density (cm^{-2})	2.33e13
ΔV_{OFF}	Correction parameter for V_{OFF} used in the calculation in the surface-potential (V)	−0.21

shows the variation in the simulated $E_{\perp,D}$ with V_{GS} at 25°C at various drain-to-source voltages.

The extracted parameters for the Poole–Frenkel (PF) current are tabulated in Table 8.3. The extracted value for the zero-electric-field barrier height for electron emission from the trap state to the continuum of states, ϕ_{PF0}, at temperature $T = 25°$C is 0.7 eV, which is in the same range of values reported in the literature [71]. Assuming the trap state to be very close to the metal Fermi level, the continuum of states is at a height equal to $\phi_{PF0} = 0.7$ eV at $T = 25°$C from the metal Fermi level. Figure 8.2 shows the variation of $\phi_{PF0,T}$, given by (8.32), with temperature. As can be noted, $\phi_{PF0,T}$ shows a 54% decrease in its value over the temperature range from 25°C to 200°C.

The extracted parameters for the Fowler–Nordheim (FN) current are tabulated in Table 8.4. To fit the model to the measured data, we set the effective barrier height used in the Fowler–Nordheim model equal to the Schottky barrier height, as these two barrier heights are expected to be in close match [71], yielding a critical barrier width of 8.5 nm. This critical electric barrier width corresponds, using (8.44), to a critical electric field of $E_c = 110.6$ MV/cm at $T = 25°$C. The variation of $\phi_{FN,T}$, given by (8.44), with temperature is shown in Fig. 8.2. As can be noted from this

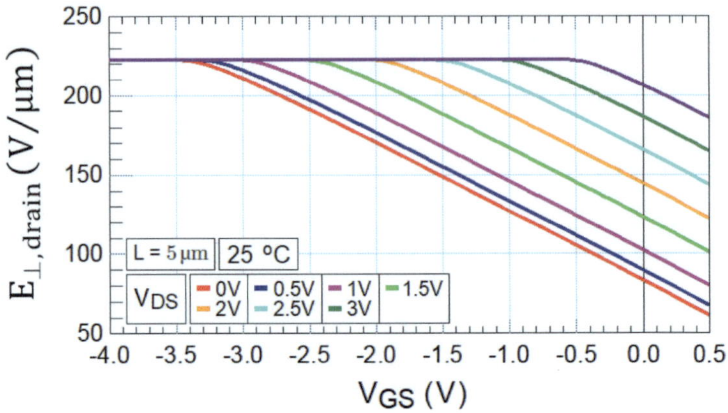

Fig. 8.4 The variation of the simulated vertical electric field at the drain side of the channel, $E_{\perp,D}$, with the gate-to-source voltage, V_{GS}, at the ambient temperature $T = 25°C$, at the drain-to-source voltages, V_{DS}, ranging from 0 V to 3 V

Table 8.3 Poole–Frankel (PF) model parameters

	Description	Value
ϕ_{PF0}	Zero-electric-field barrier height for the electron emission from the trap states to the continuum of states at 25°C (eV)	0.7
$K_{\phi_{PF}}$	Temperature parameter associated with ϕ_{PF} (eV)	−0.64
C_{PF}	The constant in the equation of J_{PF} (A V^{-1}m^{-1})	19.9e-15
κ	Parameter relating the high frequency and DC dielectric permittivity of the AlGaN barrier	0.3

figure, $\phi_{FN,T}$ shows a 3% decrease in its value over the temperature range from 25°C to 200°C.

The excellent fit of the proposed model to the measured gate current under forward-bias and reverse-bias conditions is shown in Fig. 8.3. The accuracy shown in this figure confirms that the two contributing mechanisms in gate leakage in the reverse-bias condition are the PF emission and FN tunneling mechanisms through the AlGaN barrier, as opposed to leakage current that can also occur through the AlGaN surface along the source and drain access regions, as reported in [80–82], in which case the electric field depends on the potential differences between the gate and source/drain terminals and the lengths of the access regions. The model also

Table 8.4 Fowler–Nordheim (FN) model parameters

	Description	Value
ϕ_{FN}	Effective barrier height (eV)	0.94
$K_{\phi FN}$	Temperature parameter associated with ϕ_{FN} (eV)	−48.3e-3
C_{FN}	The constant in the equation of J_{FN} (A V^{-2})	11.7e-9
d_c	Critical barrier width (m)	8.5e-9

shows the consistent modeling of the drain and gate currents. Note that both the gate and drain currents saturate at the same gate-to-source voltage, corresponding to the saturation of the vertical electric field in the AlGaN barrier, as shown in Fig. 8.4. Note further that at gate-to-source voltages less than the threshold voltage, i.e., $V_{GS} < -3$ V, $I_D = I_G$, implying, as expected, that in OFF state, the gate leakage is the only contributor to the drain current.

In order to show the contribution of the FN current, we simulated the I_G-V_{GS} characteristic for the device with gate length $L = 5\,\mu\text{m}$ at $V_{DS} = 0$ V, at various ambient temperatures, ranging from $T = 25°\text{C}$ to 200°C, disabling the FN current by setting the parameter C_{FN} in (8.41) to 0. The results are shown in Fig. 8.3a. In order to show the contribution of the PF current, we then simulated the same I_G-V_{GS} characteristic, disabling, this time, both the FN and the PF currents by setting the parameter C_{FN} and the parameter C_{PF} in (8.25) to 0. The results are also shown in Fig. 8.3a. Finally, we simulated the I_G-V_{GS} characteristic for the device with gate length $L = 5\,\mu\text{m}$ at $V_{DS} = 1.5$ V and 3 V, setting the parameter C_{FN} to 0. The results are shown in Fig. 8.3b. This figure shows that the Poole–Frenkel current becomes a main contributor to the gate current at high temperatures.

Chapter 9
Effect of Ambient Temperature on GaN Device

9.1 Introduction

The circuits designed with GaN HEMTs may be deployed in environments where temperature can vary over a range. Typical ambient temperature range for commercial electronics is 0°–70°C, while for industrial and military application, the operating temperature ranges are −40° to 85°C and −55° to 125°C, respectively. To guarantee the circuit performance within these operating temperature ranges as required by the application area, circuit simulations need to be performed at several ambient temperatures. The simulations at multiple ambient temperatures need simulation models to capture the change in device performance over the range of ambient temperatures of interest.

The modeling of effect of ambient temperature on circuit simulations gets implemented in the following way. The ambient temperature condition for the simulation is set in the circuit simulation option. For instance for the original SPICE, .TEMP option allows users to set the ambient temperature. Simulations where ambient temperature is swept to observe the change in circuit performance over ambient temperature are also allowed in simulators. The ambient temperature set by .TEMP serves as an input to the compact model.

Compact models take the input ambient temperature from the simulator. Compact models include ambient temperature in the formulations. It is important to understand that the model parameter extraction is performed at nominal temperature T_{nom}. T_{nom} is a model parameter in the ASM-HEMT model. Typically, T_{nom} is 25°C. For ambient temperature equal to T_{nom}, the model formulation ensures no effect of ambient temperature, and temperature effect gets introduced for simulation when ambient temperature is different than T_{nom}. This is achieved in ASM-HEMT with formulation implemented in manner shown in Eq. (9.1).

$$P_T = P_0 \cdot \left(1 + TP \cdot \left(\frac{T}{T_{nom}} - 1\right)\right), \tag{9.1}$$

S. Khandelwal, *Advanced SPICE Model for GaN HEMTs (ASM-HEMT)*,
https://doi.org/10.1007/978-3-030-77730-2_9

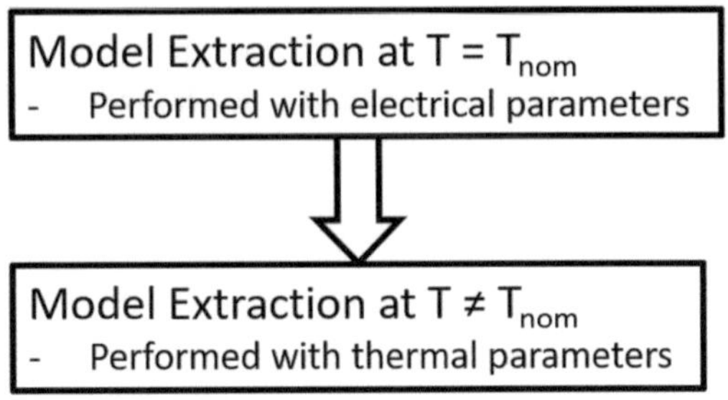

Fig. 9.1 Extraction flow process with the ASM-HEMT model for multiple ambient temperatures

where P_T is the parameter value at ambient temperature T, P_0 is the nominal parameter value, T_{nom} is the nominal temperature for parameter extraction, and TP is the model parameter to capture the temperature dependence of parameter P. When $T = T_{nom}$, the formulation ensured $P_T = P_0$. P_0 can be considered as *electrical* model parameter, and TP can be considered to be a *thermal* model parameter. The above example is one example modeling formulations to model the temperature dependence of a parameter P. Different mathematical formulations for different model parameters are typically used in compact models to accurately model its temperature dependence.

This approach is typically applied in compact models to capture the ambient temperature dependence relative to the nominal extraction temperature T_{nom}. The model extraction procedure when modeling for multiple ambient temperatures is shown in Fig. 9.1. First model extraction should be performed for nominal temperature T_{nom}. This extraction flow should use the usual electrical model parameters for the core and real device effects. Next, the model extraction for temperatures not equal to the nominal temperature should be performed using the model parameters that capture the ambient temperature dependence of electrical model parameters.

Physically, there are several device parameters that change due to the change in ambient temperature. To describe the modeling approach used in the ASM-HEMT model for all of these parameters, these are divided into core model parameters and parameters associated with real device effects. This division of model parameters is similar to one discussed in previous chapters of the book.

9.2 Temperature Dependence of Parameters Affecting Drain Current

The core model parameters capture the device electrostatics and carrier transport for an ideal device. With change in ambient temperature, electrostatics and transport change. This ambient temperature has been modeled in the ASM-HEMT model with temperature dependence of model parameters.

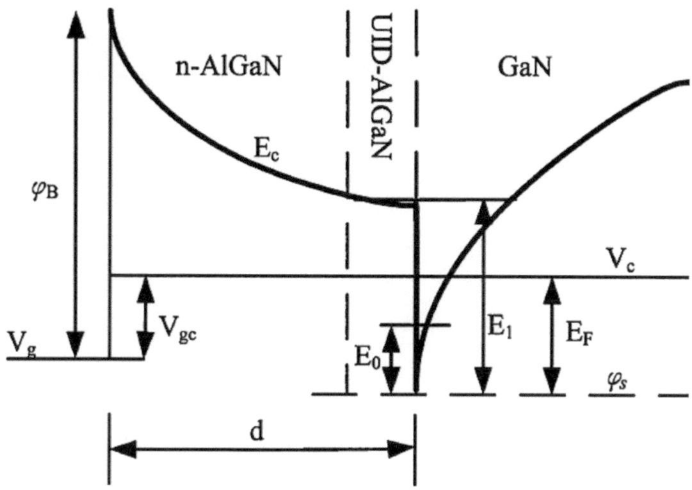

Fig. 9.2 Energy band diagram in GaN HEMTs

9.2.1 Cut-Off Voltage

The cut-off voltage V_{OFF} of GaN HEMT is the key parameter capturing the device electrostatics and is one of the core model parameters. Considering the energy band diagram shown in Fig. 9.2, V_{OFF} can be defined as

$$V_{off} = \phi_b + \Delta E_c - \frac{qN_D \cdot d_d^2}{\epsilon} - \frac{q\sigma}{\epsilon} \cdot (d_d + d_i) \,, \tag{9.2}$$

where ϕ_b is the barrier height, ΔE_c is the difference in the conduction band level between the barrier layer and the GaN layer, N_D is the doping in the barrier layer, d_d is the thickness of the doped layer of the barrier layer, d_i is the thickness of the intrinsic barrier layer, ϵ is the barrier layer dielectric permittivity, and σ is the polarization charge density. The barrier height depends on the metal work-function and electron affinity of the barrier layer. The gate metal work-function depends on temperature as described in [83]. Δ_{EC} depends on the band-gaps that depend on the temperature, the third term in Eq. (9.2) does not have a temperature dependence term, and the polarization charge σ is found to not have any significant temperature dependence. Overall, the temperature dependence of V_{off} is modeled with a linear temperature dependence in the ASM-HEMT model. The ASM-HEMT formulation is

$$V_{off}(T) = V_{off,0} - KT1 \cdot \left(\frac{T}{T_{nom}} - 1 \right), \tag{9.3}$$

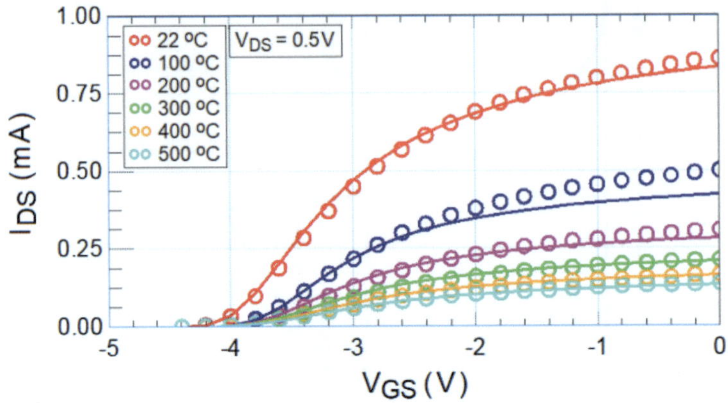

Fig. 9.3 Transfer characteristics of GaN HEMT at ambient temperature upto 500°C modeled with ASM-HEMT compact model. Drain voltage $V_d = 0.5$ V

where $V_{off}(T)$ is the value of the cut-off voltage at ambient temperature T, $V_{off,0}$ is the value of the cut-off voltage at $T = T_{nom}$, and $KT1$ is the ASM-HEMT model parameter capturing the temperature dependence of V_{off}. $KT1$ can be extracted from the slope of extracted V_{off} versus temperature plot. This linear temperature dependence is found to result in good model accuracy for a typical wide temperature range upto 500°C as shown in Fig. 9.3.

9.2.2 *Mobility*

The next core model parameter that changes with the ambient temperature is low-field carrier mobility μ_0. Mobility is the parameter that relates how fast carriers move in a semiconductor under the influence of the applied electric field. For the applied electric field E, the carrier in a semiconductor will move with velocity v given by

$$v = \mu \cdot E, \tag{9.4}$$

where μ is the mobility of carriers in the semiconductor. In the absence of any external field, the carriers move randomly due to their thermal energy. They collide with each other. As external field, the carrier drift under the influence and mobility captures an average or effective directional carrier movement in the presence of random thermal collisions. These collisions are scattering events that reduce effective carrier velocity. As the ambient temperature changes, the thermal energy of carrier and the vibration of lattice changes. This results in change in the scattering rates and in turn the carrier mobility.

Table 9.1 ASM-HEMT model parameters for low-field carrier mobility and the corresponding model parameters for modeling their temperature dependence

Low-field mobility parameter	Parameter for temperature dependence
U_0	UTE
$U_0, ACCS$	UTES
$U_0, ACCD$	UTED
$U_{0,FPx}$	UTE

The change in carrier mobility is modeled with Eq. (9.4). This non-linear dependence has been applied and shown to capture the mobility variation for a temperature range upto 500°C as shown in Fig. 9.3.

$$\mu(T) = \mu_0 \cdot \left(\frac{T}{T_{nom}}\right)^{UTE}, \tag{9.5}$$

where μ is the mobility at the ambient temperature T, μ_0 is the mobility at the nominal temperature, and UTE is the model parameter that is used to capture the temperature dependence of the mobility.

In ASM-HEMT, the mobility of source- and drain-side access regions, and channel under field-plates have their own low-field mobility parameter. Typically, these values are equal; however, they may be tuned to be slightly different for improved model accuracy. Each of these low-field model parameters also have the ambient temperature dependence included in the model. The low-field mobility parameters and their corresponding temperature dependent parameters in the ASM-HEMT model are given in Table 9.1. $U_{0,FPx}$ in Table 9.1 has $x = 1$ to 4 for the four field-plate regions in the ASM-HEMT model. As noted by the model formulations, the ambient temperature dependence of mobility is non-linear power law function in contrast to the linear ambient temperature dependence of the cut-off voltage.

9.2.3 Saturation Velocity

As the applied drain–source voltage and as a result the lateral electric field increase, the carriers move in the channel with increasingly higher velocity. However, the linear increase in the carrier velocity with electric field remains valid only till the carriers are in thermal equilibrium with the lattice. At higher electric fields, the carrier energy increases above the thermal energy and they can lose energy via optical phonon emission. As a result, the carrier velocity saturates to saturation velocity v_{sat}. This effect is modeled in the ASM-HEMT model with the model parameter $VSAT$. The optical phonon emission is dependent on the

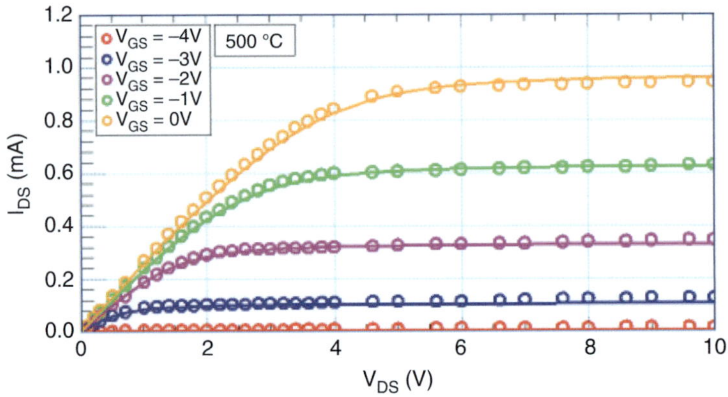

Fig. 9.4 Output characteristics of GaN HEMT at ambient temperature upto 500°C modeled with ASM-HEMT compact model

Table 9.2 ASM-HEMT model parameters for saturation velocity and the corresponding model parameters for modeling their temperature dependence

Saturation velocity parameter	Parameter for temperature dependence
VSAT	AT
VSATACCS	ATS
VSATACCD	ATD
VSAT0FPx	AT

ambient temperature, and as a result, the saturation velocity parameter has ambient temperature dependence. This is modeled with the formulation Eq. (9.6).

$$v_{sat}(T) = v_{sat0} \cdot \left(\frac{T}{T_{nom}}\right)^{AT}, \tag{9.6}$$

where v_{sat} is the saturation velocity at the ambient temperature T, v_{sat0} is the saturation velocity at the nominal temperature, and AT is the model parameter that models the ambient temperature dependence of the saturation velocity. The mathematical form of ambient temperature dependence of the saturation velocity is non-linear power law dependence. This formulation has been shown to be accurate upto 500°C as shown in Fig. 9.4 as shown in [84]. Similar to the modeling approach used for the carrier mobility, the temperature dependence for the different saturation velocity parameters in the access regions and field-plate regions has parameters for temperature dependence as shown in Table 9.2.

9.2.4 Access Region Charge Density

The two-dimensional electron gas density in the access regions is also found to be dependent on temperature. This is observed by modeling devices for a wide range of temperatures. The surface trap states are considered to be the source of two-dimensional electron gas density. As the temperature increases, the occupancy of these surface states changes causing the temperature dependence of two-dimensional electron gas density.

In the ASM-HEMT model, the following formulation is used to model the temperature dependence of two-dimensional electron density in the source and drain access regions.

$$NS0ACCS(T) = NS0ACCS\left(1 - KNS0\left(\frac{T_{dev}}{TNOM} - 1\right)\right) \tag{9.7}$$

$$NS0ACCD(T) = NS0ACCD\left(1 - KNS0\left(\frac{T_{dev}}{TNOM} - 1\right)\right). \tag{9.8}$$

9.2.5 Contact Resistance

The source and drain contacts are ohmic contacts. The resistivity of the contacts depends on the ambient temperature. As the temperature increases, the scattering rate of electrons increases. This leads to an increase in resistivity. This is modeled with the following formulations in the ASM-HEMT model for the source- and drain-side contact resistances.

$$RSC(T) = RSC\left(1 + KRSC\left(\frac{T_{dev}}{TNOM} - 1\right)\right) \tag{9.9}$$

$$RDC(T) = RSC\left(1 + KRDC\left(\frac{T_{dev}}{TNOM} - 1\right)\right). \tag{9.10}$$

The model parameters $KRSC$ and $KRDC$ can be used to capture the temperature dependence of contact resistivity. Two separate parameters are added for tuning the impact of contact resistivity at the source and drain ends of the device. Typically, these two parameters will have the same value. Two parameters are provided for fine tuning in model extraction.

9.3 Temperature Dependence of Gate Current Parameters

Gate current parameters for both forward- and reverse-bias conditions depend on ambient temperature. For the forward-bias condition, the barrier voltage is affected by the ambient temperature. This is modeled with the following formulation:

$$VBI(T) = VBI - \left(\frac{T_{dev}}{TNOM} - 1\right) KTVBI. \tag{9.11}$$

The forward ideality factor of Schottky diodes is also temperature dependent. This dependence of model with

$$NJG(T) = NJG + \left(\frac{T_{dev}}{TNOM} - 1\right) KTNJG. \tag{9.12}$$

The temperature dependence of saturation current of the diodes is modeled with the following formulation.

$$IGDIO(T) = IGDIO \cdot exp\left(KTG \cdot (T/TNOM) - 1\right). \tag{9.13}$$

9.4 Self-Heating Effect and Modeling for Multiple Ambient Temperatures

Self-heating occurs due to the power dissipation in the device. Considering the high power levels handled by GaN HEMTs are used, self-heating is an important effect to model. Self-heating is dependent on bias conditions as the power dissipated in the device is given by

$$P_d = V_{ds} \cdot I_{ds}, \tag{9.14}$$

where V_{ds} is the applied drain–source terminal voltage, and I_{ds} is the drain current. Only drain–source voltage and current are considered as typically drain–source current is significantly higher than the gate current.

The amount of local temperature rise in the device due to the power dissipation is a transient phenomenon as governed by the thermal time-constant of the device. Thermal time-constant is modeled with a combination of thermal resistance and capacitance as shown in Fig. 9.5. The thermal resistance models the static temperature rise or the temperature rise in the DC conditions. Thermal capacitance is needed in the model to capture the dynamic nature of the self-heating effect. For instance, if power is dissipated in the device for a duration smaller than the thermal constant, the local temperature of the device will not increase. This thermal network is used

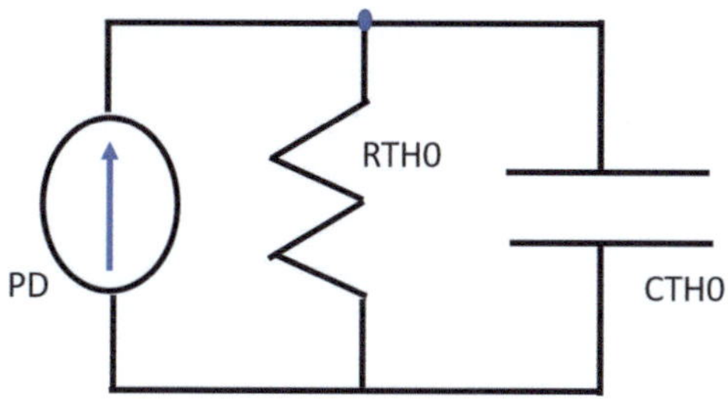

Fig. 9.5 Thermal network for modeling the self-heating effect

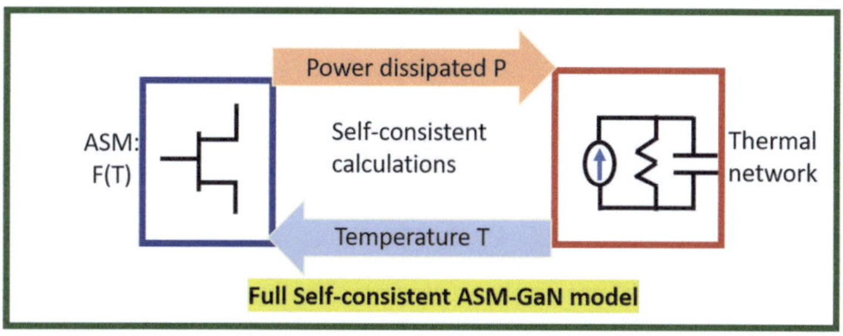

Fig. 9.6 Self-consistent model of self-heating effect in the ASM-HEMT model

in the ASM-HEMT model to capture the self-heating effect with thermal resistance RTH0 and thermal capacitance CTH0 as model parameters.

The temperature rise due to the self-heating effect is added to the local temperature to obtain the device temperature in the ASM-HEMT model.

$$T_{device} = T_{ambient} + \Delta T. \tag{9.15}$$

The local device temperature T_{device} is used in calculating the final temperature affected parameters using the formulations discussed before. T_{device} is dependent on power dissipated in the device, and power dissipation in the device is dependent T_device via the device parameters. This interdependence shown in Fig. 9.6 is self-consistently solved in the ASM-HEMT model.

There are several methods available in the literature for extraction of the thermal resistance and capacitance in the device. Two methods are discussed here. First, thermal simulations with finite element analysis can be performed by creating a three-dimensional model of the device. The observed power dissipated in the device is taken to be the heat source, and the three-dimensional device structure is divided into several small elements. Heat flow equations are solved at each point to get the thermal resistance and capacitance of the device.

Another method to extract thermal resistance and capacitance is from specific electrical measurements on the device. One of the popular methods for thermal resistance extraction is to use gate metal resistance as the thermometer. In this method, the gate resistance is first obtained for multiple ambient temperatures. During this measurement, the device is off and no power is dissipated in the device. After this calibration step, the device is powered and gate resistance is measured again. By combining the two data-sets, the local rise in temperature with power is extracted. The slope of temperature rise with power is the thermal resistance of the device. The thermal capacitance can be extracted from the measured frequency response of the output conductance. After the frequency corresponding to the thermal time-constant, the output conductance increases.

Chapter 10
Noise Models

10.1 Flicker Noise

Assuming the probability of tunneling of an electron to a trap site decreases exponentially with the distance of the trap from the channel, we get power spectral density (PSD) of the mean square fluctuations in the number of occupied traps as

$$S_N(f) = N_t \cdot \frac{kT}{N_{P_t} f^E F}, \tag{10.1}$$

where Nt is the concentration of traps, k is the Boltzmann constant, and T denotes the temperature in Kelvin. The N_{P_t} factor comes from tunneling probability, and EF is a modeling parameter. The drain current noise PSD is then given by

$$S_{if}(f) = \frac{S_N(f)}{WL^2} \int_{0_L} \left(\frac{dI_{DS}}{dN_t} \right)^2 \cdot dx. \tag{10.2}$$

Denoting the 2DEG charge in the channel as Q, we can write the fluctuation in the carrier density caused by the fluctuation in the trapped electron density as

$$R = \frac{1}{q} \frac{dQ}{dN_t}. \tag{10.3}$$

R varies with applied biases, and the calculation of the exact expression for R is shown in the appendix. The drain current IDS is given as

$$I_{DS} = \mu W(-Q_{ch}(x)) \frac{\partial \psi}{\partial x}. \tag{10.4}$$

S. Khandelwal, *Advanced SPICE Model for GaN HEMTs (ASM-HEMT)*,
https://doi.org/10.1007/978-3-030-77730-2_10

Our unified flicker noise model takes both the theories of mobility fluctuation and carrier number fluctuation into account. The power spectral density (PSD) of the mean square fluctuations in the number of occupied traps is given as [85, 86]

$$S_N(f) = N_t \frac{k_B T_{dev}}{N_{P_t} f^{EF}}, \tag{10.5}$$

where N_t is the concentration of traps (in cm^{-2}) and f is the frequency of operation. The N_{P_t} factor comes from tunneling probability and EF is a modeling parameter. This in turn gives the drain current noise PSD as

$$S_{if}(f) = \frac{S_N(f)}{WL^2} \int_0^L \left(\frac{\partial I_{DS}}{\partial N_t}\right)^2 dx. \tag{10.6}$$

Also,

$$\partial I_{DS} = \mu W(-Q_{ch}) \frac{\partial \psi}{\partial x} \left[\frac{1}{-Q} \frac{\partial(-Q)}{\partial N_t} \pm \frac{1}{\mu} \frac{\partial \mu}{N_t} \right] \partial N_t \tag{10.7}$$

$$\partial I_{DS} = I_{DS} \left[\frac{qR}{-Q_{ch}} \pm \frac{1}{\mu} \frac{\partial \mu}{\partial N_t} \right] \partial N_t, \tag{10.8}$$

where $R = \frac{1}{q} \frac{\partial(-Q_{ch})}{\partial N_t}$. The + or − sign depends on whether a trap is neutral or charged when filled, which will increase or decrease mobility. Using (10.6) and (10.8), we get the drain current noise PSD as [87]

$$\begin{aligned} S_{if}(f) = \frac{k_B T}{WL^2 f^{EF}} \frac{I_{DS}^2 K_r}{C_g^2} \Bigg[& NOIAV_{th}C_g \left(\frac{1}{Q_{ch,d}} - \frac{1}{Q_{ch,s}} \right) \\ & + (NOIA + NOIBV_{th}C_g) ln \left(\frac{Q_{ch,d}}{Q_{ch,s}} \right) \qquad (10.9) \\ & + (NOIB + NOICV_{th}C_g)(-Q_{ch,d} + Q_{ch,s}) \\ & + \frac{NOIC}{2} \left(Q_{ch,d}^2 - Q_{ch,s}^2 \right) \Bigg], \qquad (10.10) \end{aligned}$$

where $K_r = L/\left[(V_{go} - \psi_m + V_{th})(\psi_d - \psi_s)\right]$, $Q_{ch,(d/s)} = qC_g\left(V_{go} - \psi_{(d/s)}\right)$, and $NOIA$, $NIOIB$, and $NOIC$ are the parameters that include terms like N_{P_t} and N_t.

The developed model is validated against both TCAD and experiments. The model validation with TCAD is shown in Fig. 10.1. Experiments show that applying a large gate voltage induces an increase in the number of traps getting filled and hence increases flicker noise. However, with increasing gate voltage, the increase in drain current overpowers the increase in the noise resulting in an overall decreasing

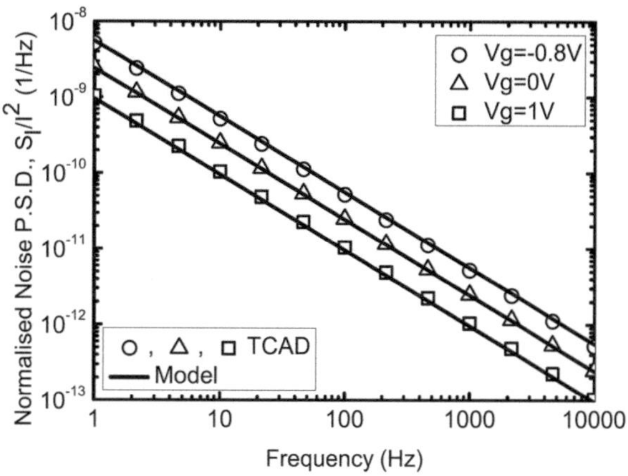

Fig. 10.1 Model validation with TCAD simulation of flicker noise

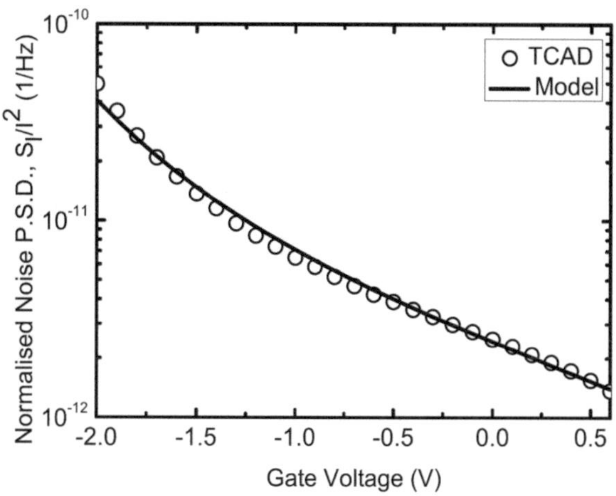

Fig. 10.2 Model validation with TCAD simulation of flicker noise

curve for the normalized noise PSD. This can be seen in the variation of the normalized drain current noise PSD with gate bias as shown in Fig. 10.2, which illustrates the excellent matching between the TCAD data and the ASM-HEMT model.

The comparison of the model and the measured flicker noise for a 0.7 μm GaN device is shown in Fig. 10.3. The increasing trend of the flicker noise with increasing gate voltage is clearly shown in Fig. 10.4. The figure uses measured data from a 1 μm GaN HEMT.

10.2 Thermal Noise

Thermal noise is caused by the thermal excitation of charge carriers. The multiple collision processes that a charged carrier undergoes are widely regarded as the source of this type of noise. The thermal noise model proposed by Klaassen

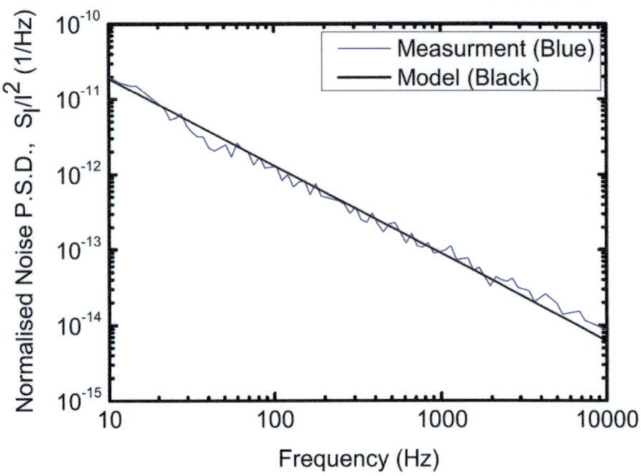

Fig. 10.3 Model validation with measurements of flicker noise

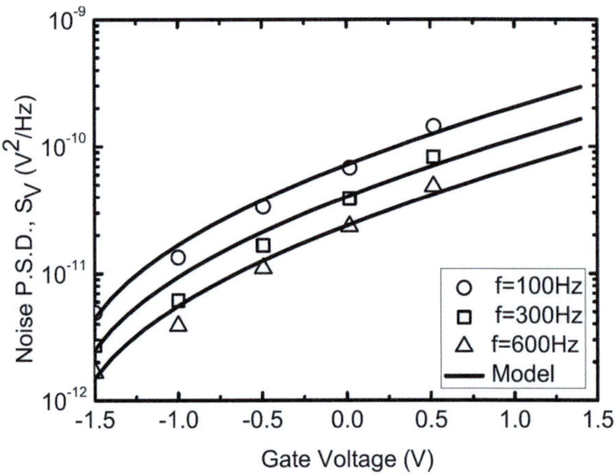

Fig. 10.4 Model validation with measurements of flicker noise

and Prins [12] forms the basis of most of the thermal noise models of today, albeit with modifications to include various effects. With our model, we aim at a computationally simple, yet highly tunable model for thermal noise in HEMT devices. For this, we have used the equation for the power spectral density (PSD) of the channel thermal noise proposed by Klaassen and Prins (KP), as the basis of our derivations. Writing the KP model in terms of the surface-potential, we get the noise PSD as

$$S_{it} = \frac{4k_B T_{dev}}{I_D L_{eff}^2} \int_{\psi_s}^{\psi_d} g^2(\psi) d\psi, \tag{10.11}$$

where $g(\psi) = \mu_{eff,sat} W(qn_s)$.

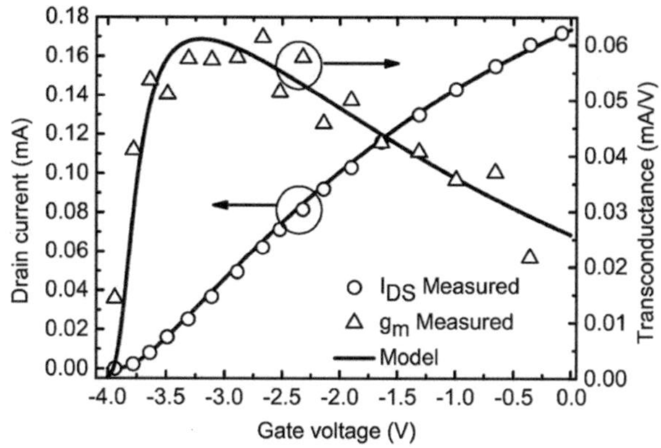

Fig. 10.5 Model and measurements for current and trans-conductance

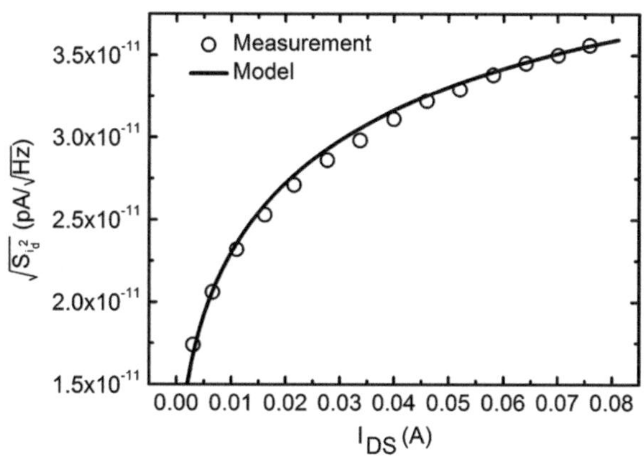

Fig. 10.6 Model and measurements for current and trans-conductance

The final noise PSD can therefore be written as [88]

$$S_{it} = \frac{4k_B T_{dev}}{I_D L_{eff}^2} \left(\mu_{eff,sat} W q C_g\right)^2 \left(V_{go}^2 \psi_{ds} + \frac{\psi_d^3 - \psi_s^3}{3} - V_{go} \left(\psi_d^2 - \psi_s^2\right) \right). \tag{10.12}$$

A plot showing the variation of the drain current and the trans-conductance with the gate bias for a $2 \times 100\,\mu\text{m}$ GaN device with a channel length of 100 nm is shown in Fig. 10.5. Figure 10.6 shows the PSD as a function of the same device. Increasing drain current implies an increasing number of charge carriers, which in turn results in more collisions, and hence, more noise is generated in the device as demonstrated by the increasing trend of the noise PSD with increasing drain current in this figure.

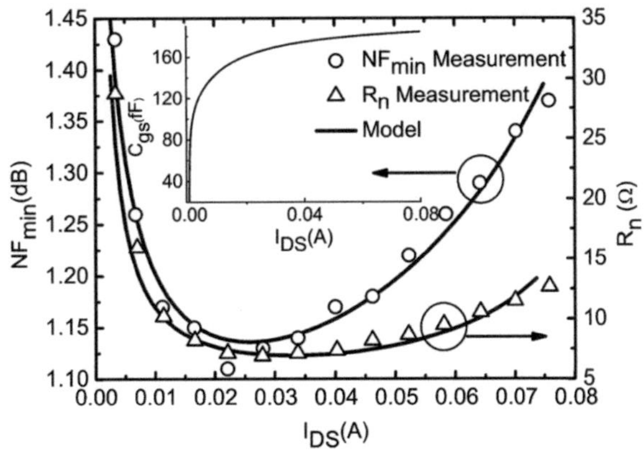

Fig. 10.7 Model and measurements for current and trans-conductance

However, the variation in trans-conductance shown in Fig. 10.5 also needs to be considered for minimum noise figure. The variation of the minimum noise figure with drain current is shown in Fig. 10.7 for the aforementioned device, along with the equivalent noise resistance, and shows good fit with our model accuracy.

Chapter 11
Parameter Extraction in ASM-HEMT Model

11.1 GaN RF Device Extraction Flow

This section describes the radio-frequency GaN parameter extraction flow with ASM-HEMT model. I-V, and S-parameters measurements are used for extraction, and large-signal non-linear simulations are used for validation, and fine-tuning of the model.

11.1.1 DC Parameter Extraction

The DC model parameter extraction flow for the ASM-GaN-HEMT model can be summarized in the figure given below. The following steps describe the flow:

1. Set physical parameters in the model such as L, W, *NF*, *TBAR*, *LSG*, and *LDG*. These parameters are generally available to the device technologist.
2. First the extraction is focused on getting the linear VD condition parameters. From linear drain current bias conditions, i.e., V_D 50–100 mV $ID-VG$ characteristics, extract the parameters such as *VOFF* and *NFACTOR*. *VOFF* is the cut-off voltage and is a very important parameter. A rough estimate of *VOFF* is the *VG* values in the $ID-VG$ plot in linear scale from which *ID* starts to rise. This rough value can be fine-tuned as one moves along in the parameter extraction flow for best fits. *NFACTOR* controls the sub-threshold slope of the device, and this can be extracted by fitting the model to linear VD condition $ID-VG$ characteristics on logarithm scale.
3. After *VOFF* and *NFACTOR* carrier low-field mobility $U0$ and mobility vertical-field dependence parameters *UA* and *UB* should be extracted from linear VD condition $ID-VG$ characteristics. Trans-conductance GM and its derivatives GM' and GM'' in the linear VD condition can be accurately modeled using $U0$,

S. Khandelwal, *Advanced SPICE Model for GaN HEMTs (ASM-HEMT)*,
https://doi.org/10.1007/978-3-030-77730-2_11

UA, and *UB* parameters. It is important to note here that the series resistance also affects linear $ID - VG$. It is helpful to keep the values of series resistance parameters RSC, $NS0ACCS, D$, *MEXPACSS*, *D* to reasonable values for this step. The starting values for series resistance parameters (which will be fine-tuned) can be from special measurements like TLM structures or simply close to previously extracted devices.
4. With linear VD condition fitting done, the parameters from the high VD conditions should be extracted. As done above, first we focus on $ID - VG$ characteristics for high VD conditions and extract the sub-Voff or low current region parameters. Due to the drain-induced barrier lowering effect, the cut-off voltage in high VD condition decreases, and this can be modeled with the extraction of DIBL parameters $ETA0$ and *VDSCALE*. The sub-threshold slope also degrades for high VD conditions, and this can be modeled with parameter *CDSCD* in the ASM-GaN-HEMT model.
5. Next, above-Voff fitting for high VD $ID - VG$ curved is done. The key parameters for this fitting are the velocity saturation parameters $VSAT$, channel-length modulation parameter (*LAMBDA*), and the non-linear series resistance parameters $NS0ACCS, S$, *MEXPACCS,D* and $U0ACCS, D$. These parameters can be extracted by fitting high VD $ID - VG$ in linear scale. $VSAT$ is best extracted from intermediate current levels for which GM in high VD conditions is still rising with *VG*. The non-linear series resistance causes GM to drop with *VG*, and the associated parameters should be tuned to fit this region. Self-heating will also have an impact on high current and high drain voltage regions, and the thermal resistance of the device should be set to a reasonable value from special measurements or TCAD simulations or from previous knowledge. Accurate modeling of trans-conductance for different VD conditions can be achieved with the parameters indicated in this step. It is good to go back to linear *IDVG* after this step and do a very fine adjustments for further improved fits.
6. Next, the output characteristics $ID - VD$ should be looked at. With *IDVG* fitted under different VD conditions, $IDVD$ should be already accurately modeled. Parameters can be fine-tuned to get improvement, if required.

The above steps complete DC parameter extraction flow for the ASM-GaN-HEMT model at room temperature. For modeling DC I-Vs at other temperatures, temperature scaling equations on the key model parameters are implemented in the model. Using the temperature parameters DC I-Vs at multiple temperatures can be modeled (Fig. 11.1).

11.1.2 S-Parameter Model

After DC parameters, S-parameter can be modeled with the model. For this, the parasitic components need to be accounted for, and a sub-circuit should be built around the model to represent all the parasitic capacitances and inductances.

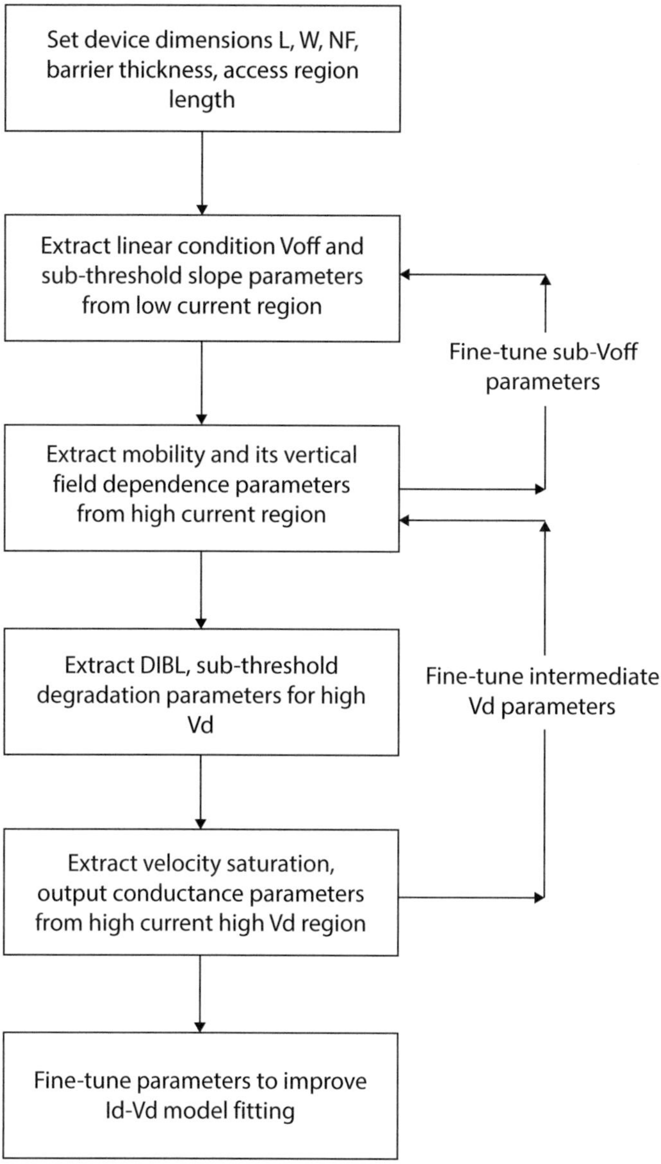

Fig. 11.1 Parameter extraction procedure

The parasitic elements associated with the actual layout and the position of the measurement reference planes need to be included with the intrinsic ASM GaN model. The complete model can be visualized as shown in Fig. 11.2, consisting of three levels: level-1 is the intrinsic ASM GaN model, level-2 is the model of electrode inductances, resistances, and capacitances associated with metal lines of

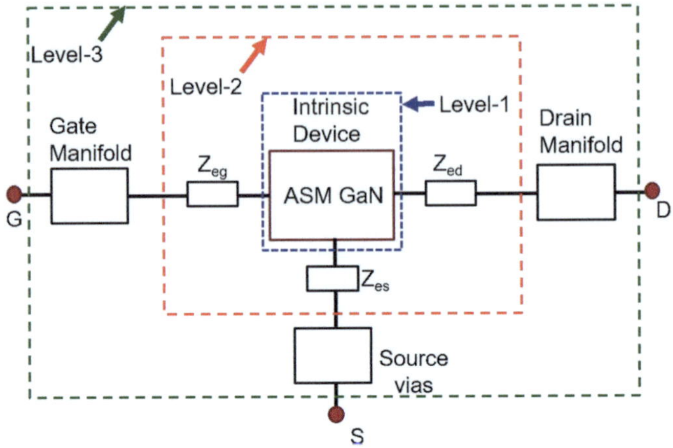

Fig. 11.2 ASM GaN model intrinsic device with parasitic for full radio-frequency model

the electrodes represented by Z_{ex} with x = g, d, or s, and level-3 are the parasitic elements for the source via gate and drain manifolds. Level-3 parasitic effect can be either modeled with a lumped element network, or the S2P files from EM simulations can be directly used for modeling this region. Level-2 parasitic elements can be extracted from the standard extraction flow in the literature.

Using the correct parasitic networks, the gate–source, gate–drain, and drain–source capacitances can be observed from measurements and the model from imaginary parts of Y-parameters. The parameters $CGSO$, *ADOSI*, *BDOSI*, and *VDSATCV* can be used to tune the C_{gs} versus V_{gs} behavior at different $Vds's$. $CGSO$ should be tuned to set the value of C_{gs} for V_{gs} biases below the cut-off voltage. *ADOSI* and *BDOSI* can tune the rate of C_{gs} increase with V_{gs}.

Next, C_{gd} versus V_{ds} characteristics, the $CGDL$ parameter can be tuned to control the decrease in C_{gd} with the increase in V_{ds}. For fitting the C_{ds} parameters, *CDSO*, $CJ0$, AJ, and DJ can be used. The effect of gate resistance seen at high frequency can be modeled with gate resistance parameters $RSHG$ and XGW.

The small signals g_m and g_{ds} for different bias conditions directly originate from already modeled I-V behavior and come out reasonably close to the measured data for low frequency. At higher frequency, the frequency dependence in g_m and g_{ds} coming from self-heating can be tuned with $CTH0$ parameter.

After S-parameter modeling at multiple DC bias points, large signal RF modeling of input power sweep curves and load/source-pull measurements can be attained. One of the important regions of I-V to model for large signal performance is the knee region in the I-V region. This region is controlled mainly by the access region resistance parameters and the smoothing parameters DELTA, MEXPACCS, and MEXPACCD.

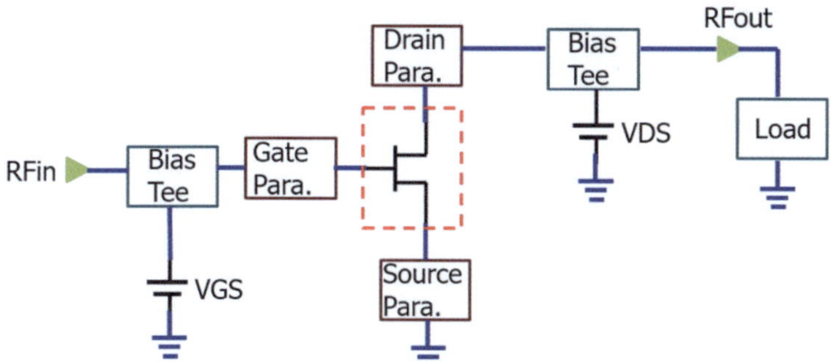

Fig. 11.3 Large signal simulation setup with the ASM-HEMT model

11.1.3 Non-Linear Large Signal Model

The large signal simulation setup for ASM-HEMT is shown in Fig. 11.3. The transistor symbol represents the ASM-HEMT model. RFin in this figure represents the input RF power. V_{GS} is applied to the device through the input bias tee. At the input gate, parasitic input network block is also shown. This parasitic network includes the gate manifold model and the additional parasitic elements extracted following the process described above. Equivalent to the input side, the output or the drain side also contains bias tee and the parasitic network. The drain parasitic network includes the drain manifold model and other parasitic elements extracted in the previous step. RFout is the output signal from the device that is observed at the applied load. The simulation setup also contains the model for source vias. Source via model can be based on appropriate equivalent circuit or the via can be modeled with electromagnetic simulations of the structure.

The setup described in Fig. 11.3 can be used for multiple bias conditions, frequency, source, and load impedance conditions. The previous extraction flow covered a full range of bias conditions, and the resulting ASM-HEMT model is multi-bias model. The large signal device behavior is the result of its I-V and C-V behavior. During the extraction flow, we modeled I-V and C-V accurately. Large signal modeling and simulations do not require any parameter extraction steps. These simulations are for model validation and fine tuning. Sample model results after large signal for tuned load conditions are shown in Fig. 11.4.

Load-pull simulations can also be simulated with the setup shown in Fig. 11.3. The load impedance points for fundamental and harmonic frequencies should be set to appropriate values as in the measurements for these simulations. Typically, these values can be directly read into the simulations from the measured data files of an appropriate format. An example simulation results for load-pull simulations comparing model results and simulations are shown in Fig. 11.5.

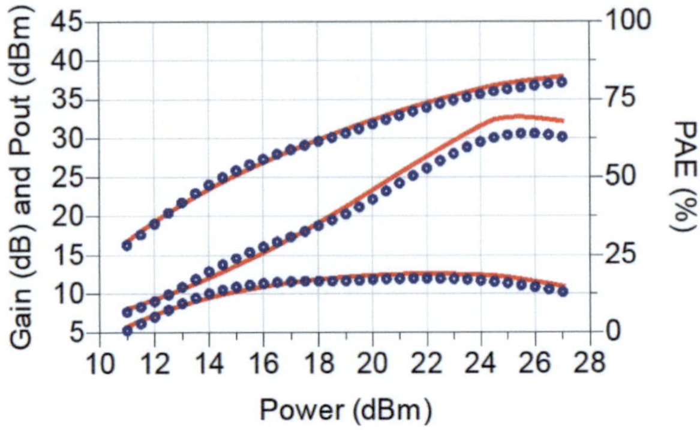

Fig. 11.4 Large signal model results with ASM-HEMT for tuned load conditions

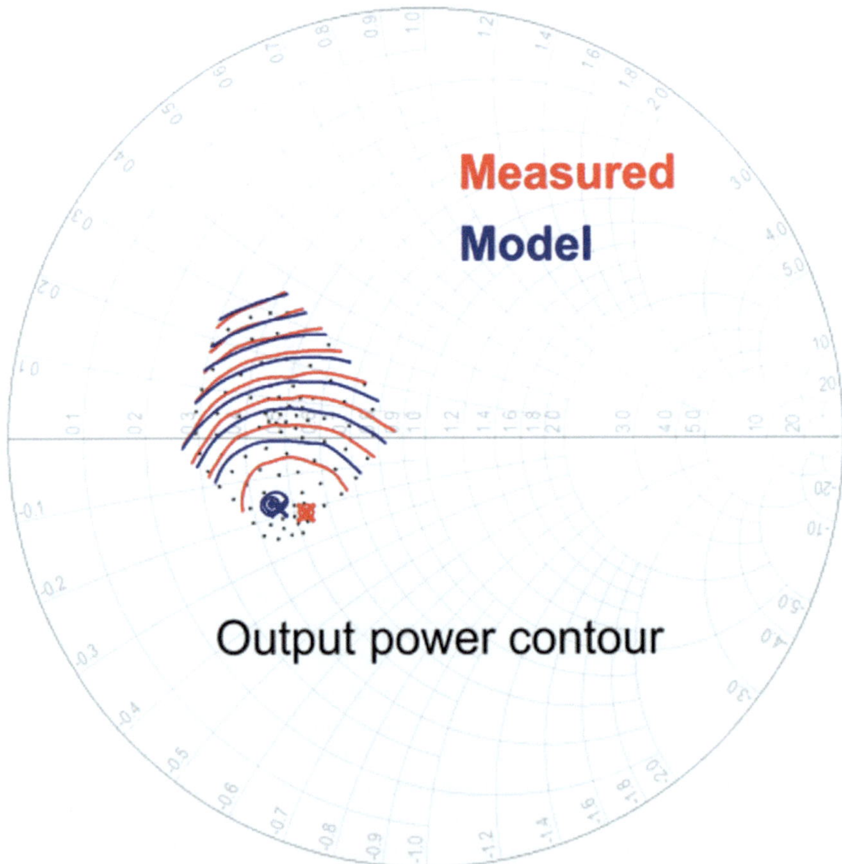

Fig. 11.5 Load-pull simulations with the ASM-HEMT model

11.2 GaN Power Device Extraction Flow

The extraction flow for power device follows the similar general strategy as discussed for the RF GaN HEMTs. However, there are key differences. Typically, I-V and C-V data are taken for modeling the power devices. I-V behavior is measured with semiconductor analyzers, and C-V is measured with impedance analyzers. Using these measurements and information about the device geometry, an accurate ASM-HEMT model can be developed for GaN power devices.

Some GaN power devices are combined with silicon FETs to obtain a normally off device operation. Normally off GaN switches comprising of a low voltage Si MOSFET and a high voltage GaN HEMT in cascode configuration have demonstrated to have superior performance and reliability [89]. This technology has also been qualified to automotive standards [90]. To exploit the full potential of this device platform, advanced circuit simulations with accurate models of the switches are needed. The available GaN power device models [91, 92] either use empirical non-linear functions [91] or are based on very simple general transistor model formulations [92]. These approaches do not account for key physical device effects in GaN power device such as non-linear access region resistance and complex voltage-dependent field-plate capacitances. Furthermore, these models have decoupled or independent formulations for I-V and C-V behavior of the device, making the I-V and C-V simulations inconsistent with each other. Recently, a physics-based modeling approach was presented [93]. However, complex voltage-dependent field-plate capacitances were not modeled in [93]. While physics-based models have enabled advanced simulation capability in other power device technologies [94], current GaN power device models are lacking on this front. This chapter addresses this need by adapting proven physics-based radio-frequency (RF) GaN device model for GaN power switches. We apply the ASM-HEMT [95] compact model, accepted by the consortium Compact Model Coalition (CMC) [96], to advance the state-of-the-art of GaN power device modeling. ASM-HEMT has been shown to model radio-frequency (RF) GaN technology successfully [40, 47, 97]. This chapter extends the RF model to handle multiple field-plates in a power HEMT and combines it with the silicon MOSFET model for a complete model of the 650 V cascode GaN switch.

In such a scenario, the ASM-HEMT model can be used for modeling GaN HEMT of the normally on power device. ASM-HEMT can be combined with the silicon FET model for developing full model of the power device. This is discussed in the extraction flow below.

11.2.1 I-V Parameter Extraction

The I-V extraction flow starts with the first setting or configuring the model for power device. This includes setting the model switches and geometry information.

Fig. 11.6 Model development flow for modeling the full 650 V cascode power device

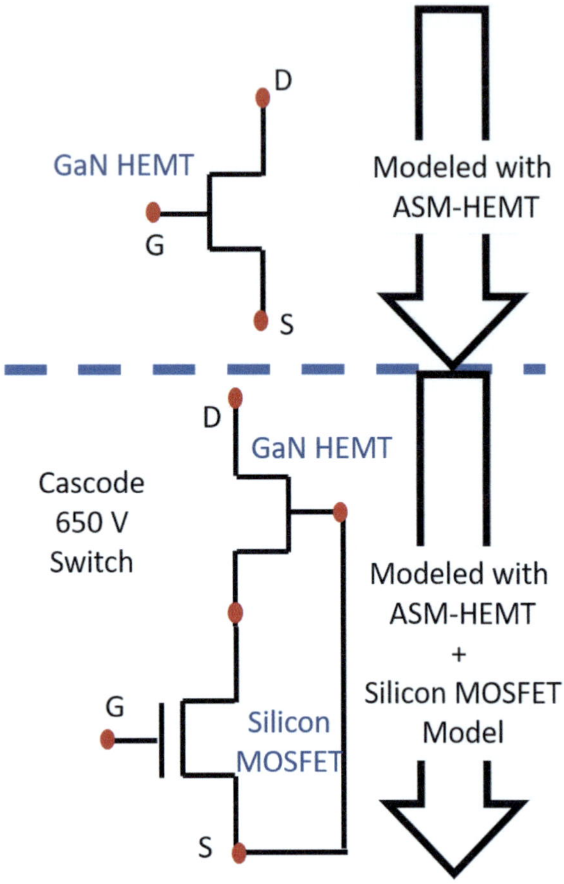

GaN power devices also have a larger number of field-plates as compared to RF GaN HEMTs, and the field-plate switches for the device should be properly configured at the start of the extraction flow. The I-V extraction flow for normally on power device can follow the same extraction flow as was discussed for the RF GaN HEMT. So this is not discussed further. For normally off GaN power switches created with the use of silicon FETs, the extraction flow for the complete model needs some adjustments. For such a device, the extraction flow is shown in Fig. 11.6. First, the GaN HEMT part of the complete power switch is independently measured and modeled. The I-V extraction flow for the GaN HEMT can follow the same extraction strategy as for the RF GaN HEMT. Example results obtained for commercial power GaN HEMT obtained after following the extraction flow are shown in Figs. 11.7 and 11.8. Next, the C-V of the GaN HEMT part of the power device should be modeled to develop a complete model for the GaN HEMT part of the power switch. The extraction flow for C-V for power devices is discussed in the next sub-section. After the full model for the GaN HEMT is complete, this model can be combined with the model of the

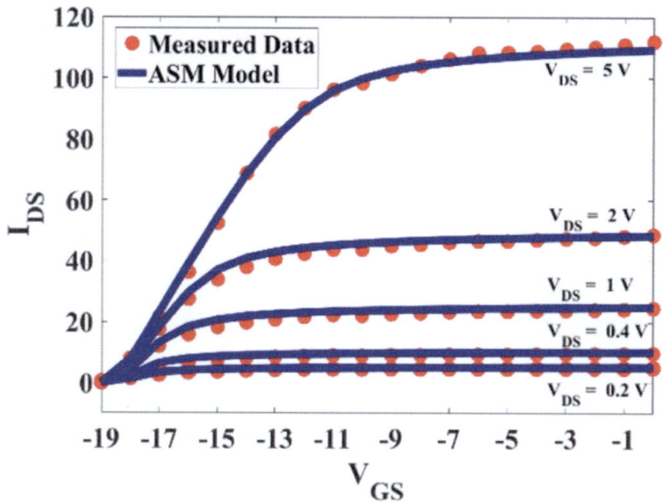

Fig. 11.7 Transfer characteristics of the GaN HEMT in a cascode device is accurately modeled with a combination of ASM-HEMT and silicon FET model

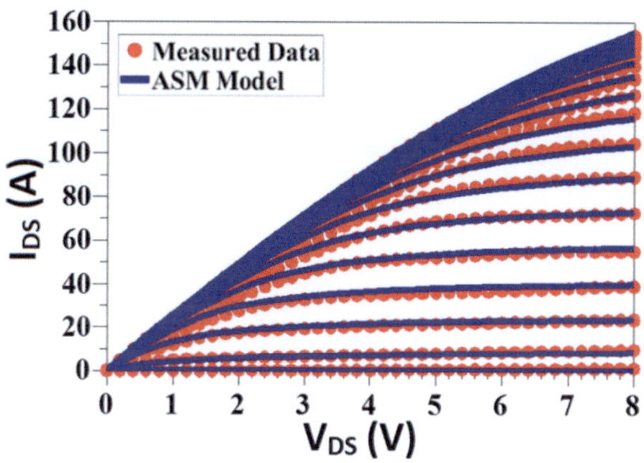

Fig. 11.8 Output characteristics of the GaN HEMT in a cascode device is accurately modeled with a combination of ASM-HEMT and silicon FET model

silicon FET used for the normally off operation in configuration and it is used in the complete power switch, for instance cascode configuration. Usually commercial silicon FETs are used for this purpose, and the models for these are available. ASM-HEMT-based GaN model and the silicon FET model can then be combined as shown in Fig. 11.9. After combining, the full power device can be simulated and checked against the measured I-V of the device. Example model results for complete power device obtained after such combined model are shown in Figs. 11.10 and 11.11.

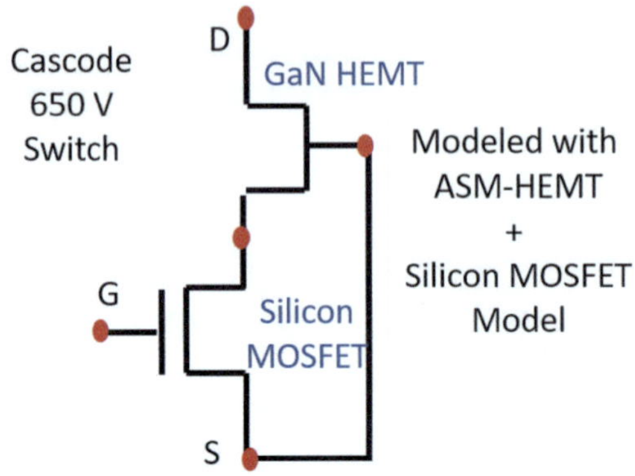

Fig. 11.9 Cascode power device model combining silicon FET model with ASM-HEMT-based GaN model

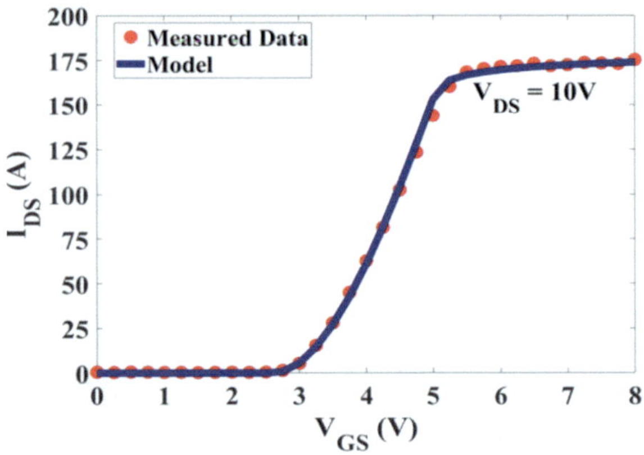

Fig. 11.10 Transfer characteristics of the complete 650 V cascode power switch is accurately modeled with the ASM-HEMT model

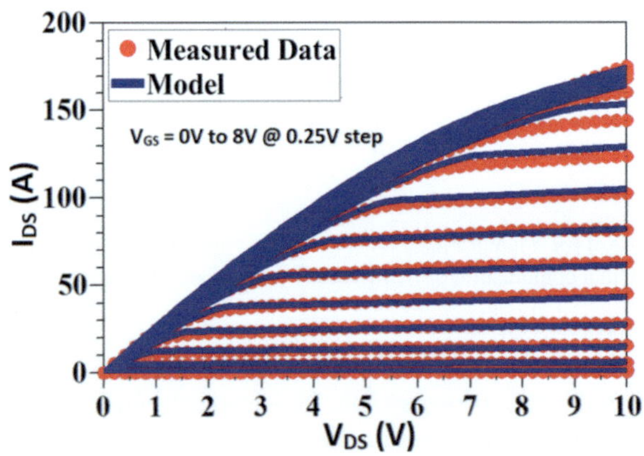

Fig. 11.11 Output characteristics of the complete 650 V cascode power switch is accurately modeled with the ASM-HEMT model

11.2.2 Capacitance Modeling

The C-V characteristics including the input capacitance C_{iss}, the output capacitance C_{oss}, and the reverse transfer capacitance C_{rss} versus drain–source voltage are

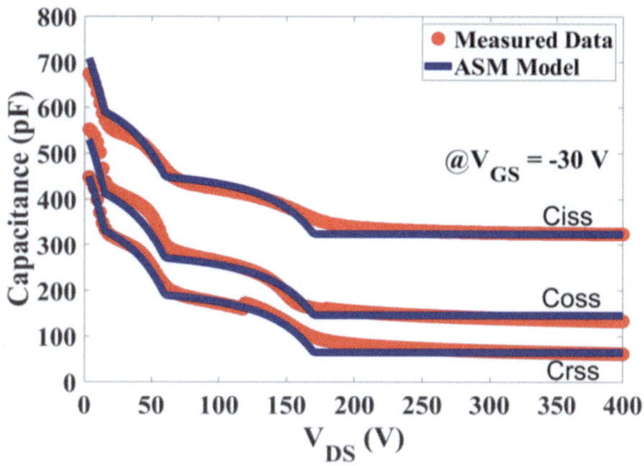

Fig. 11.12 Input C_{iss}, output C_{oss}, and cross capacitance C_{rss} of GaN HEMT of cascode switch modeled with a combination of ASM-HEMT and silicon FET model. The complex bias dependence of capacitance due to the presence of field-plates is accurately modeled. The jump in C_{rss} around $V_{ds} \approx 125$ V is due to range switching in the measuring instrument

accurately modeled as shown in Fig. 11.12. In the ASM-HEMT model, C-V behavior of the device is modeled with the analytical formulations of the terminal charges of the device. The formulation for gate charge Q_g is shown in Eq. (4.19). This formalism of modeling the capacitance with terminal charges has been shown to be physical with better circuit convergence under transient simulations [98]. Nevertheless, capacitance can be simulated from the model with standard AC simulation setup for C-V. The total terminal charge also includes the field-plate region charges in the ASM-HEMT model. In the GaN HEMT in this study, there are three gate-connected field-plates. The charges in these regions are calculated with Eq. (4.19) where L_i and $C_{g,i}$ are the length and input capacitance per unit area of the i-th field-plate. The total gate charge is given by Eq. (11.1).

$$Q_{g,total} = Q_g + \sum_{i=1}^{i=3} Q_{gfp,i}. \tag{11.1}$$

The presence of field-plates in the HEMT causes a complex variation in C_{rss} versus V_{ds}. C_{rss} keeps decreasing in a stepped fashion as V_{ds} is increased to $V_{ds} \approx 200$ V and then decreases slowly thereafter (see Fig. 11.12). As V_{ds} is decreased below 180 V, the third field-plate region turns on, causing the capacitance to increase from the values at $V_{ds} > 180$ V. Further decrease in V_{ds} turns on the second field-plate around $V_{ds} \approx 80$ V, and then the first field-plate next to the gate at $V_{ds} \approx 30$ V turns on. As field-plate regions turn on, they contribute to the capacitance of the device as seen in Fig. 11.12. Accurate model for complex bias-dependent C_{rss} has been obtained for the first time for power GaN HEMTs. All three field-plate regions in the device have been accurately modeled. The charge modulation in the field-plate regions also reflects in the bias dependence seen in C_{iss} and C_{oss}. A good accuracy is achieved in model for all the three capacitances.

After combining the ASM-HEMT-based GaN power device model and the silicon FET model, accurate C-V behavior for the complete power switch can be

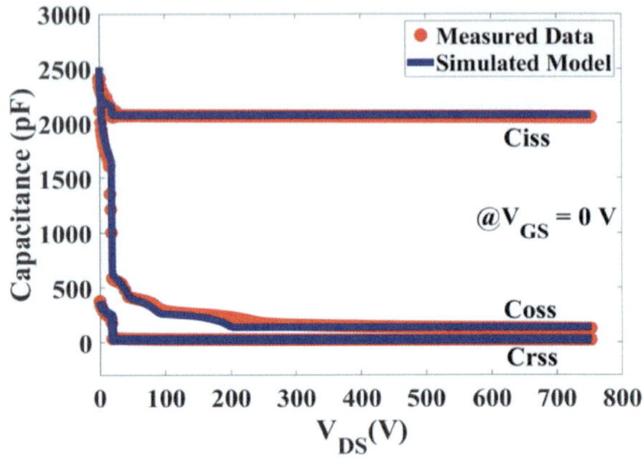

Fig. 11.13 Input C_{iss}, output C_{oss}, and cross capacitance C_{rss} of GaN HEMT of cascode switch modeled with the ASM-HEMT model. The complex bias dependence of capacitance due to the presence of field-plates is accurately modeled

obtained. An example is shown in Fig. 11.13. Usually, only fine tuning of the C-V parameters is required. The need for tuning is mainly due to the manufacturing variations in the devices, and the model used for silicon FET may not be for the exact copy of the device that is used in the cascode device.

11.2.3 Modeling the Package Parasitics

In order to model a packaged device completely, we need to consider the inside picture of a packaged device. A packaged device consists of a die containing the intrinsic FET device and a package surrounding it with bond wires connecting the die to the external pins of the package. This combination of a package material, bond wire, and external pins can be modeled as a lumped network as shown in Fig. 11.14. This package-level modeling has been found to be quite useful in detailed analysis of the high frequency switching performance of the GaN power devices [99]. There are a total of five extrinsic (L_g, C_{pg}, L_s, L_d, C_{pd}) and three intrinsic elements (RGC, RSC, RDC) in the lumped network that model the gate side, the source side, and the drain side of the equivalent model shown in Fig. 11.14.

The two intrinsic elements RSC and RDC have already been modeled in the I-V extraction flow. In order to extract the remaining five extrinsic elements of the package and the intrinsic model element RGC, the OFF-state S-parameter from frequency from 0.05 MHz to 600 MHz, and setting the gate-to-source and drain-to-source voltages to 0 V (in particular, $V_{G_{ext}S_{ext}} = V_{D_{ext}S_{ext}} = 0\,\text{V}$). Under this condition, the device is in OFF state. Under this condition, we can approximate the model in Fig. 11.9 to the one in Fig. 11.15 and the three OFF-state capacitance model parameters CGDO, CGSO, and CDSO are the only elements from the intrinsic FET device that contribute to the topology.

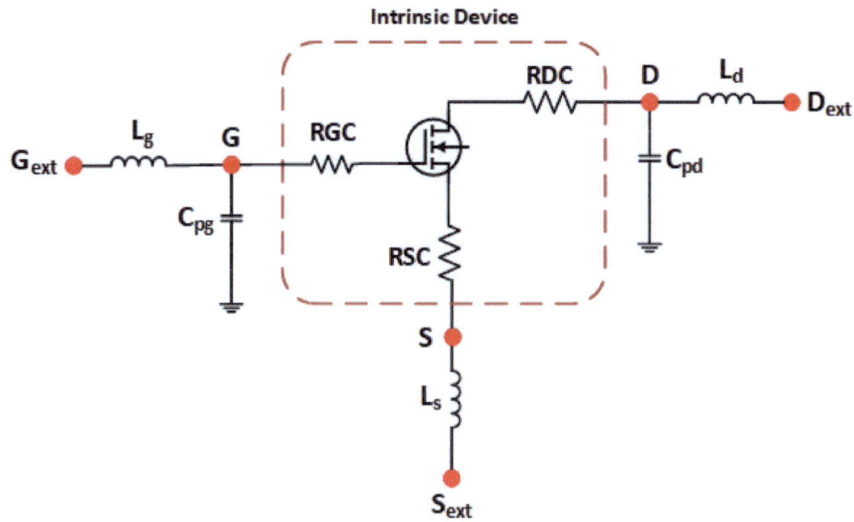

Fig. 11.14 Schematic showing the package parasitic components of a power GaN device

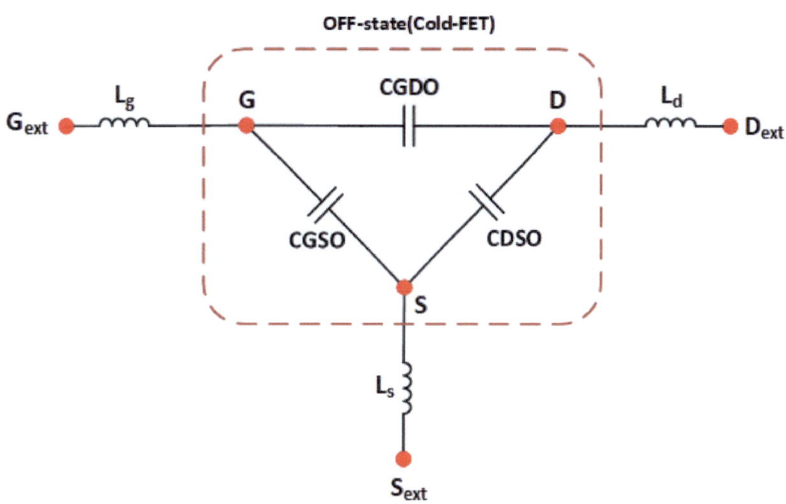

Fig. 11.15 OFF-state equivalent model for the power device

The parameters L_g, L_s, and L_d of Fig. 11.15 can easily be extracted from the imaginary parts of the Z-parameters:

$$\mathrm{Im}(z_{11}) = \omega(L_g + L_s) - \frac{1}{\omega(\mathrm{CGSO} + \frac{\mathrm{CGDO}\cdot\mathrm{CDSO}}{\mathrm{CGDO}+\mathrm{CDSO}})}, \tag{11.2a}$$

$$\mathrm{Im}(z_{21/12}) = \omega(L_g + L_d) - \frac{1}{\omega(\mathrm{CGDO} + \frac{\mathrm{CGSO}\cdot\mathrm{CDSO}}{\mathrm{CGSO}+\mathrm{CDSO}})}, \tag{11.2b}$$

$$\mathrm{Im}(z_{22}) = \omega(L_d + L_s) - \frac{1}{\omega(\mathrm{CDSO} + \frac{\mathrm{CGSO}\cdot\mathrm{CGDO}}{\mathrm{CGSO}+\mathrm{CGDO}})}. \tag{11.2c}$$

As can be noted from the measurement results of Figs. 11.16, 11.17, 11.18, and 11.19, the imaginary parts of the Z-parameters exhibit resonance at a particular frequency. As can be noted from the above equations, the resonance in $\mathrm{Im}(z_{11})$ is governed by a combination of L_g and L_s, the resonance in $\mathrm{Im}(z_{22})$ by a combination of L_d and L_s, and the resonance in $\mathrm{Im}(z_{21/12})$ by a combination of L_g and L_d. We have confirmed these dependencies by simulation. However, as is shown in Fig. 11.15, L_g is the key element affecting $\mathrm{Im}(z_{11})$, L_d is the key element affecting $\mathrm{Im}(z_{22})$, and L_s is the key element affecting $\mathrm{Im}(z_{21/12})$.

The parameters that are left to extract are C_{pg}, C_{pd}, and the gate resistance RGC. We have found that setting these parameters to zero, while extracting the other parameters, leads to a good agreement between the model and the measurement.

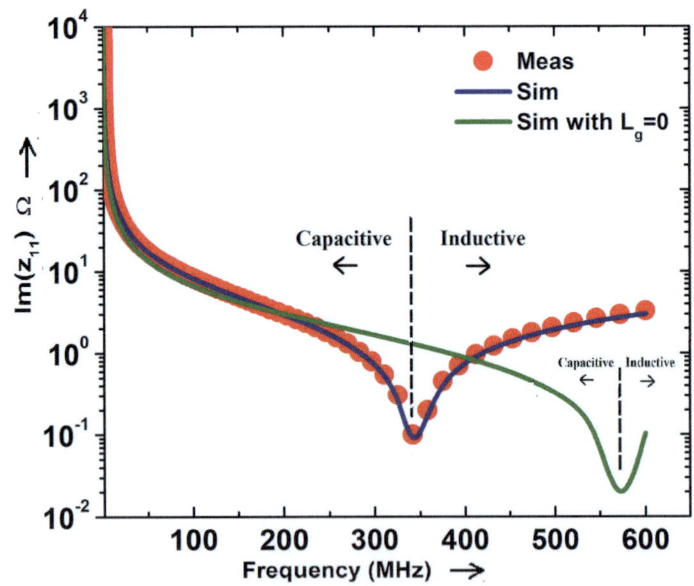

Fig. 11.16 $\mathrm{Im}\,(z_{11})$

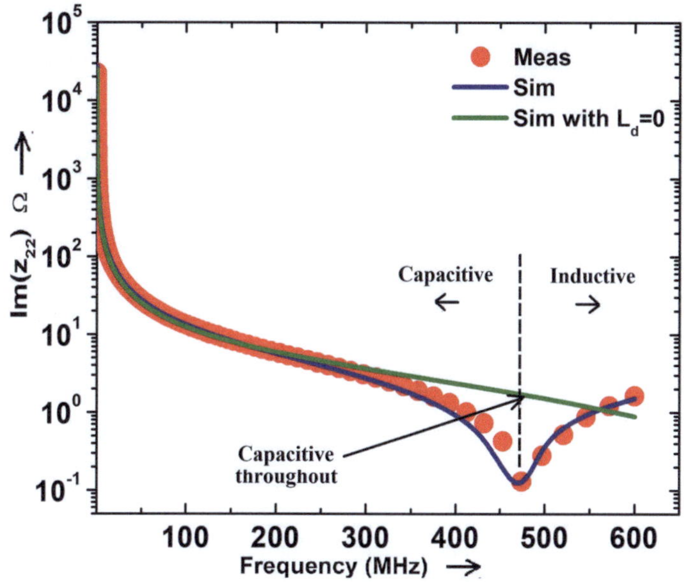

Fig. 11.17 Im (z_{22})

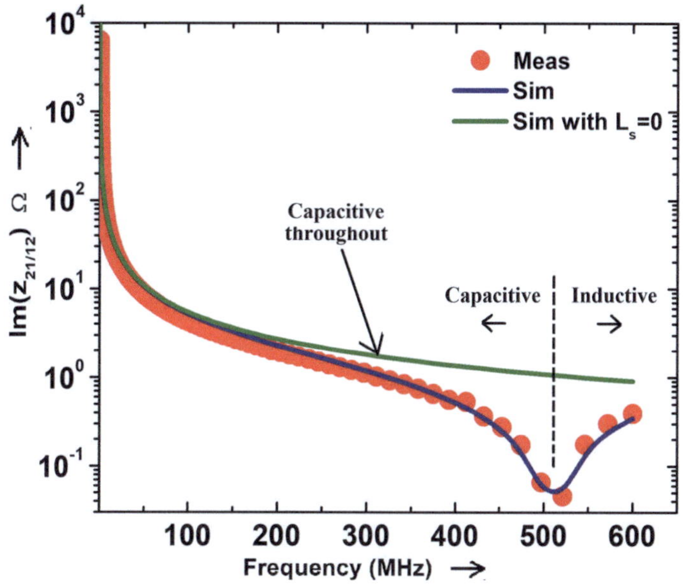

Fig. 11.18 Im $(z_{21/12})$

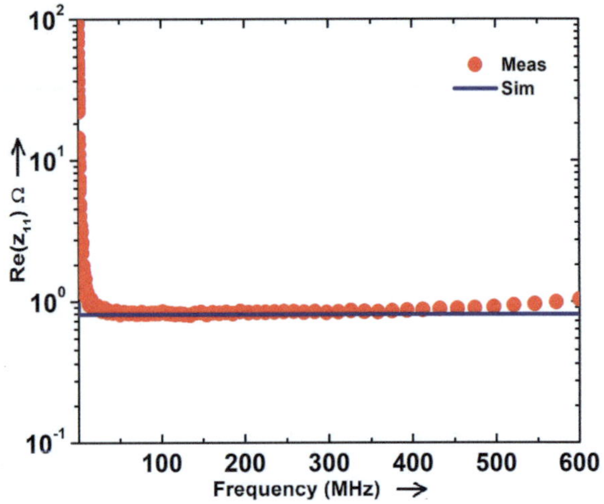

Fig. 11.19 Rem (z_{11})

The parameter RGC, on the other hand, can be extracted from $\text{Re}(z_{11})$, which is given by (see Fig. 11.9 with the internal transistor replaced by Fig. 11.15)

$$\text{Re}(z_{11}) = \text{RGC} + \text{RSC} + R_{accs}\,. \tag{11.3}$$

Hence, RGC can be extracted by tuning its value to fit the measured $\text{Re}(z_{11})$ (see Fig. 11.19).

11.3 Model Quality Tests

This section discusses the benchmark model quality tests on the ASM-HEMT model. Model has successfully passed Gummel symmetry, harmonic balance simulation, AC symmetry, and reciprocity tests. Results of these tests are given below.

11.3.1 Gummel Symmetry Test Results

The Gummel symmetry test is used to test the correctness of the model with respect to drain–source symmetry. For GaN HEMTs, it is common to use different source and drain gate lengths, which make the full device asymmetric. The intrinsic device region is, however, still symmetric, and the compact model for the intrinsic device should pass the Gummel symmetry test [100, 101].

The results are shown in Figs. 11.20, 11.21, 11.22, and 11.23.

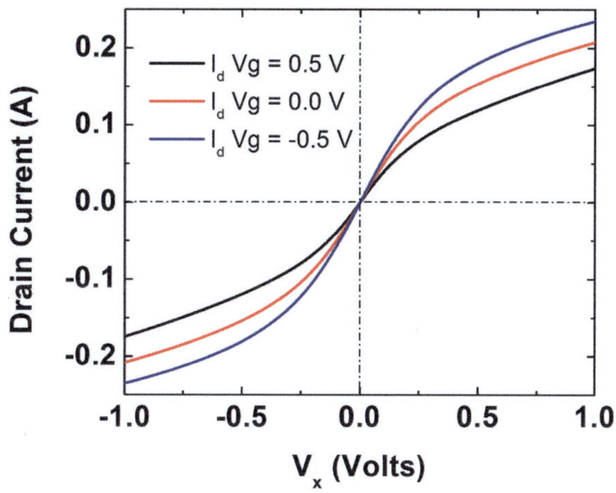

Fig. 11.20 Drain current vs. V_x

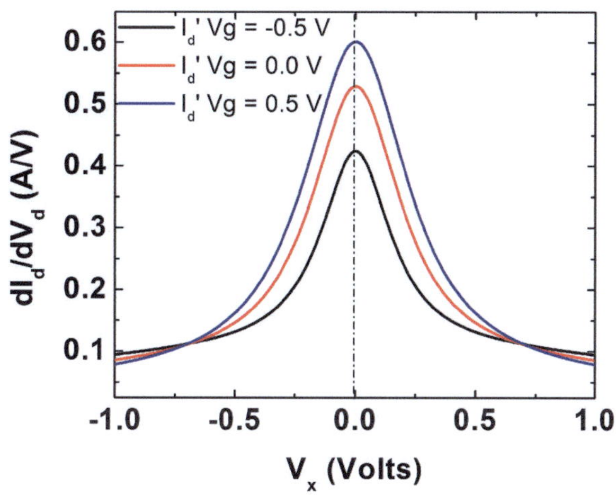

Fig. 11.21 First derivative of I_d vs. V_x

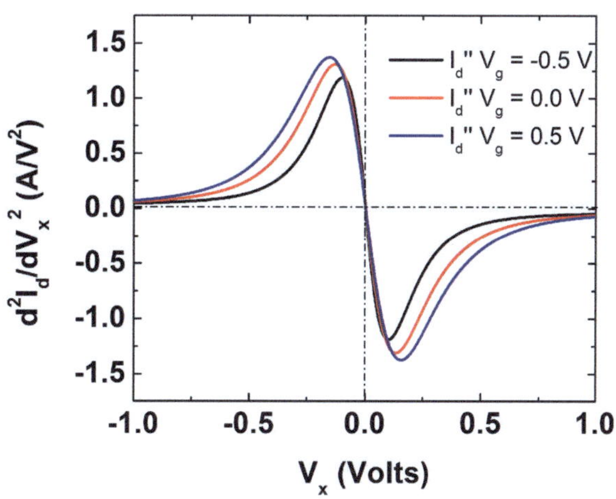

Fig. 11.22 Second derivative of I_d vs. V_x

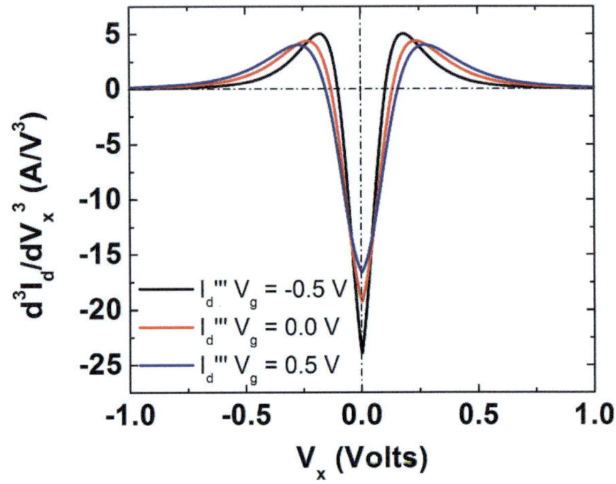

Fig. 11.23 Third derivative of I_d vs. V_x

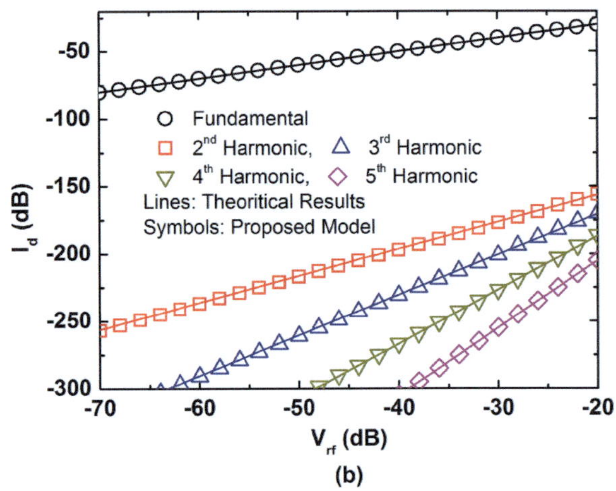

Fig. 11.24 Single tone harmonic balance simulation result

11.3.2 Harmonic Balance Simulation Results

Distortion analysis is important for evaluation of RF circuit performance, and correct model behavior for distortion analysis can be tested by the harmonic balance simulations [100, 101].

Figures 11.24 and 11.25 shows results of harmonic balance simulation results from ASM-HEMT model up-to fifth order.

11.3.3 AC Symmetry Test Results

In addition to the current expression, the charge and hence capacitance expressions in the model should also satisfy the symmetry condition. This is also known as AC symmetry test. It is tested using two expressions [100, 101].

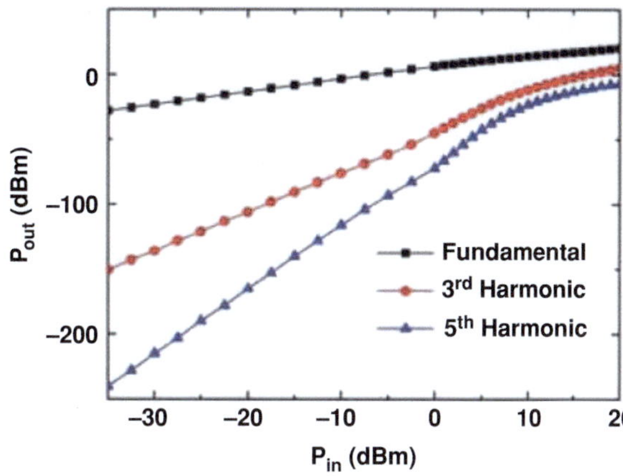

Fig. 11.25 Two tone harmonic balance simulation result

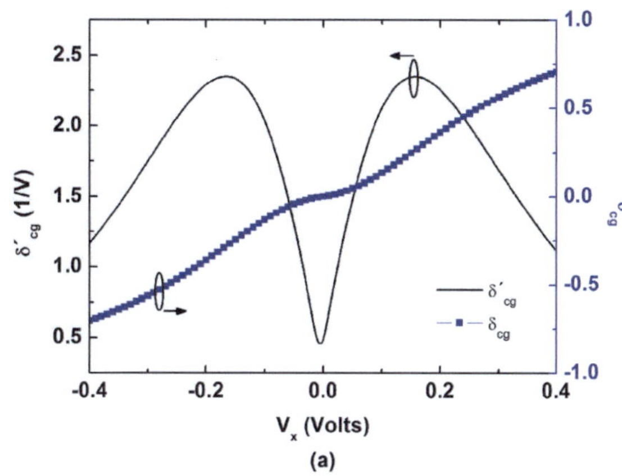

Fig. 11.26 δ_{cg} and $\delta_{cg}^{'}$ vs. V_x

$$\delta_{cg} = \frac{C_{GS} - C_{GD}}{C_{GS} + C_{GD}} \tag{11.4}$$

$$\delta_{csd} = \frac{C_{SS} - C_{DD}}{C_{SS} + C_{DD}}. \tag{11.5}$$

These functions and their derivatives are shown in Figs. 11.26 and 11.27.

11.3.4 Reciprocity Test Results

Another important test for the correct behavior of the capacitance model is the reciprocity of capacitances at zero drain–source voltage (V_{ds}). In general, the capacitance between the two nodes i and j of the device is nonreciprocal, i.e., $C_{ij} = C_{ji}$. However, $C_{ij} = C_{ji}$ at $V_{ds} = 0$, and the model should mimic this behavior [100, 101] (Figs. 11.28 and 11.29).

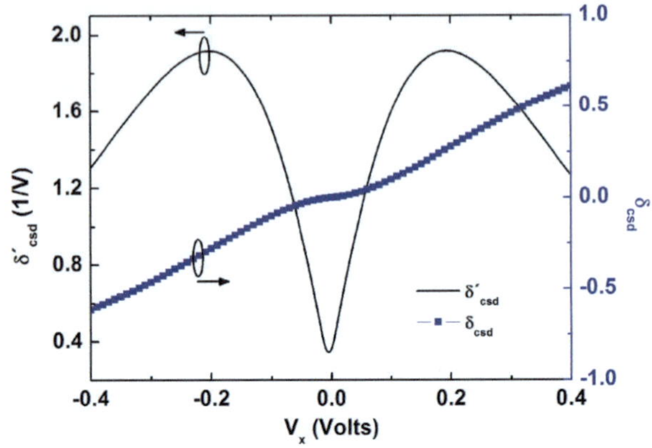

Fig. 11.27 δ_{csd} and δ'_{csd} vs. V_x

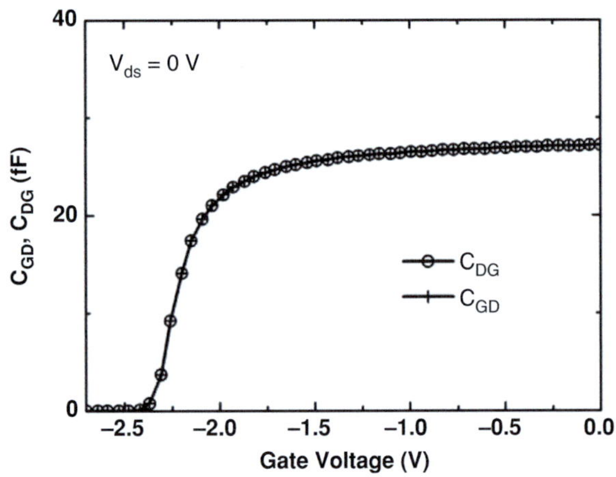

Fig. 11.28 C_{GD}, C_{DG} vs. gate voltage at $V_{DS} = 0$

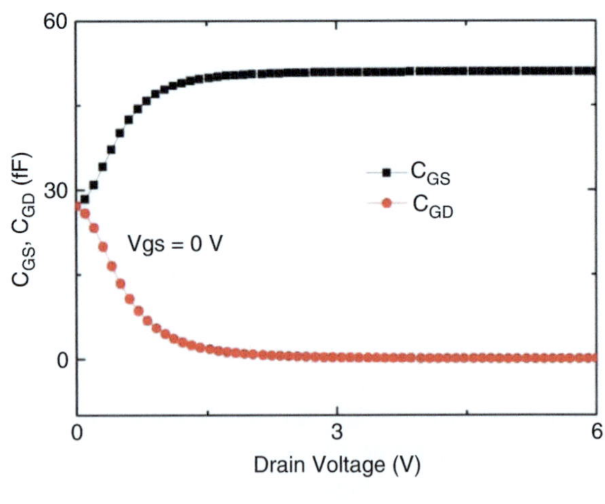

Fig. 11.29 C_{GS} and C_{GD} vs. drain voltage at $V_{DS} = 0$

Chapter 12
Advance Simulations with ASM-HEMT Model

12.1 Manufacturing Variation and RF Device Stability Analysis

Non-linear RF models for PA applications need to capture the device current and capacitance accurately. ASM-HEMT with physical foundations of the model is suited for such capabilities. The ASM-HEMT model has parameters that are directly related to the manufacturing process such as channel length (L), channel width (W), barrier thickness (T_{bar}), field-plate dimensions of the GaN device, etc. The ASM-HEMT model formulations are physically derived mathematical functions that contain the manufacturing process related dimensions as model parameters. For instance, the drain current formulation of this model is Eq. (12.1). In Eq. (12.1), one can note the parameters L and W. $C_g = \varepsilon / T_{bar}$ contains the manufacturing process parameter T_bar. In addition to the device geometry parameters, the electrical parameters such as cut-off voltage V_{off}, carrier mobility μ are also present in the drain current formulation of the ASM-HEMT model. These electrical parameters also change with the manufacturing process variations. Similar to the drain current, the terminal charge formulations of the ASM-HEMT model that are used to quantify the capacitances of the device are physically derived mathematical functions that contain device geometry and electrical parameters. For instance, the gate charge formulation is shown in Eq. (12.2). These formulations also contain geometry and electrical parameters that can vary with manufacturing process variations. Another critical property of the ASM-HEMT model in order to study the effect of manufacturing variations is that the I-V and C-V behavior of the device is modeled with a self-consistent set of geometry and electrical parameters. The model parameters and the surface-potential of the device in the I-V and C-V formulations are the same. As a result, any manufacturing variation causing change in I-V will also have causation in C-V in a physically meaningful manner in the ASM-HEMT model. This property of the model ensures that correlation between I-V and C-V is modeled physically. Owing to these properties, the ASM-HEMT

S. Khandelwal, *Advanced SPICE Model for GaN HEMTs (ASM-HEMT)*,
https://doi.org/10.1007/978-3-030-77730-2_12

model can serve as a tool to study the impact of manufacturing variations. In this chapter, we use this to analyze the non-linear stability of the device in the presence of manufacturing variations.

$$I_{ds} = \frac{\mu_{eff}}{\sqrt{1+\theta_{sat}^2 \psi_{ds}^2}} \frac{W}{L} C_g N_f \left[V_{go} - \left(\frac{\psi_s + \psi_d}{2} \right) + \frac{K_B T}{q} \right] \times \psi_{ds} \left(1 + \lambda V_{ds}\right) \tag{12.1}$$

$$\begin{aligned} Q_g = \frac{C_g L W}{\left(V_{go} - \psi_m + V_{th}\right)} & \left[V_{go}^2 + (1/3) \left(\psi_d^2 + \psi_s^2 + \psi_d \psi_s \right) \right. \\ & \left. - V_{go} \left(\psi_d + \psi_s - V_{th}\right) - V_{th} \psi_m \right]. \end{aligned} \tag{12.2}$$

The methodology for non-linear stability analysis within a commercial circuit simulator and its implementation within the Python programming language is discussed. Then we introduce the effect of manufacturing variations to analyze its impact on non-linear stability.

Traditional open-loop stability analysis methods rely on the system of interest being of a single-input single-output (SISO) nature. Applying open-loop techniques becomes increasingly more difficult in multiple-input multiple-output (MIMO) systems due to the inclusion of numerous feedback loops. For RF and mm-Wave comprised circuits, the components exhibit non-ideal input and output impedances, which forces its nature to be treated as a MIMO system, thus leading to the need for more advanced stability analysis methods. During the past several decades, there has been rigorous research [102–105] done in order to analyze the non-linear large signal stability behavior of microwave circuits. Jugo et al. [106] presented a new methodology for the analysis of large signal stability in early 2001. The method relies on obtaining the closed-loop frequency response, $H_{cl}^n(jw)$, of the circuit, which is linearized about its steady-state regime. Theory states that the closed-loop system is unstable when its closed-loop transfer function exhibits poles in the complex right-half-plane.

Figure 12.1 illustrates a test-bed schematic that has been used for the non-linear stability analysis of the extracted ASM-HEMT model. The non-linear RF model is driven by a periodic continuous-wave excitation at a pulsation $\omega_{x\gamma}$, and a harmonic balance (HB) simulation is performed. Connected to the gate terminal is a small-signal current source (SRC3). The small signal from the current source mixes with the large signal that is driving the device under test (DUT). It is to be noted here that it is important to perform the simulation over a frequency range that exceeds the bandwidth of the device. This is to ensure that no cases of instability are missed. Furthermore, it is also important to perform the analysis down to DC (0 Hz) to account for any low-frequency instabilities that may be present in the biasing networks. The mixing of the large and small signals forms a large signal small-signal analysis; thus, several transfer impedances with varying frequency translations are obtained from the circuit simulator, which can be expressed by the following:

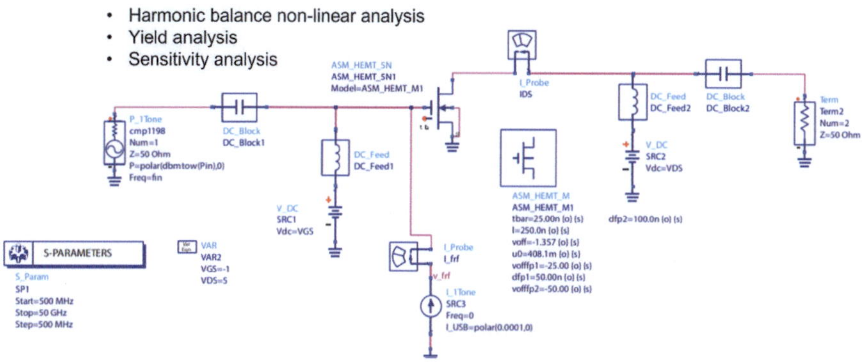

Fig. 12.1 Schematic of the test-bed used for the non-linear stability, sensitivity, and yield analyses

$$Z^n_{\alpha\beta}(j\omega) = \frac{V_\alpha(j\omega + jn\omega_{x\gamma})}{I_\beta(j\omega)} \quad n \in \mathbb{Z}, \tag{12.3}$$

where α is the node 'v_frf' in Fig. 12.1 and β is the node at which the small-signal current source is connected (gate terminal in Fig. 12.1).

Equation (12.3) takes on the form of a Harmonic Transfer Function (HTF); therefore, the circuit from Fig. 12.1 is treated as a MIMO system, and all the resultant HTFs that are obtained from the large signal small-signal HB simulation are collected in a corresponding vector, given as

$$Z_{\alpha\beta}(j\omega) = \begin{bmatrix} \vdots \\ Z^0_{\alpha\beta}(j\omega) \\ Z^1_{\alpha\beta}(j\omega) \\ \vdots \\ Z^n_{\alpha\beta}(j\omega). \end{bmatrix}$$

This provides the necessary frequency response function (FRF) that serves as the input data of the custom Python script in order to obtain the poles and zeroes of the MIMO system at different frequencies as needed for stability analysis. The custom Python script is discussed in the next section.

12.1.1 Custom Python Script

At present, closed-loop stability analysis methods have not been integrated into commercial circuit simulator tools; therefore, the analysis must be performed as a post-processing step. In order to perform the required computations to achieve the

Fig. 12.2 Simplified block diagram illustrating the incorporation of the Python script to ascertain the pole-zero plot

pole-zero maps, a custom Python script has been developed, which can be directly called upon from within the circuit simulator tool. Figure 12.2 depicts a simplified block diagram of the methodology.

Equation (12.4) is a mathematical representation of how the developed Python script in this chapter is functioning in order to obtain the pole-zeroes of the system.

$$Z^n_{\alpha\beta}(j\omega) \xrightarrow{\text{Python Script}} H^n_{cl}(jw) = \frac{\prod_{i=1}^{M}\left(jw - z_i^n\right)}{\prod_{i=1}^{N}\left(jw - p_i\right)}, \tag{12.4}$$

where p_i and z_i^n are the poles and zeros of the system's closed-loop transfer function, $H^n_{cl}(jw)$, respectively.

Olivi et al. [107] utilize a projection methodology in order to perform stable interpolation and extrapolation of FRF data. Cooman et al. [108] adopt this approach in order to perform closed-loop stability analysis. In this chapter, we make use of these efforts in order to quantify the pole-zero maps of the MIMO system in Fig. 12.1. The FRF is split into two parts: stable and unstable. This is done through projection onto the orthogonal basis of stable and unstable functions. The circuit is then considered to be unstable if the unstable basis function contains a significant portion of the FRF. In order to further discuss the projection methodology, we first introduce the notion of Hardy spaces. Traditional theory [109, 110] states that the Hardy space, $H^2(\mathbb{C}^+)$, is a subspace of the Hilbert space $L^2(j\mathbb{R})$ that allows for the decomposition given as

$$L^2(j\mathbb{R}) = H^2(\mathbb{C}^+) \oplus H^2(\mathbb{C}^-), \tag{12.5}$$

where $j\mathbb{R}$ denotes the imaginary axis.

The decomposition of (12.5) postulates that any square integrable function on $j\mathbb{R}$ decomposes uniquely as the sum of traces on the imaginary axis: an analytical function in the right-half-plane and another in the left-half-plane. Therefore, it becomes feasible to project onto $H^2(\mathbb{C}^+)$ to derive the stable part of the function, while the projection onto the orthogonal complement $H^2(\mathbb{C}^-)$ defines the unstable

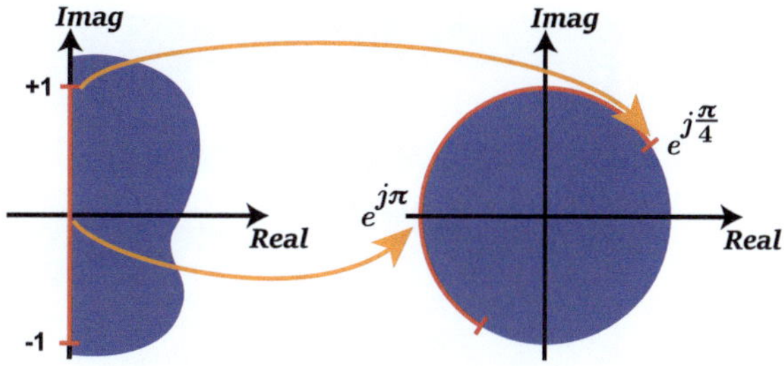

Fig. 12.3 Simplified illustration of the Möbius transformation used in the Python script to map the complex plane to the unit circle

part of the function. By assuming the output of the circuit simulator $Z_{\alpha\beta}(j\omega) \in L^2(j\mathbb{R})$, $Z_{\alpha\beta}(j\omega)$ can be expressed as the following:

$$\begin{aligned} &Z_{\alpha\beta}(j\omega) \\ &= Z_{\alpha\beta}^{\text{stable}}(jw) + Z_{\alpha\beta}^{\text{unstable}}(jw) \iff Z_{\alpha\beta}^{\text{stable}} \in H^2(\mathbb{C}^+) \;\&\; Z_{\alpha\beta}^{\text{unstable}} \in H^2(\mathbb{C}^-). \end{aligned} \tag{12.6}$$

In order to compute the stable and unstable functions, $Z_{\alpha\beta}(j\omega)$ is projected onto the basis of $H^2(\mathbb{C}^+)$ (stable) and $H^2(\mathbb{C}^-)$ (unstable), and this can be expressed by the following:

$$Z_{\alpha\beta}^{\text{stable}}(jw) = P_{H^2(\mathbb{C}^+)}\{Z_{\alpha\beta}(j\omega)\} \qquad Z_{\alpha\beta}^{\text{unstable}}(jw) = P_{H^2(\mathbb{C}^-)}\{Z_{\alpha\beta}(j\omega)\}. \tag{12.7}$$

The Python script makes use of a Möbius transformation in order to map the frequency axis to the unit circle such that the complex right-half-plane is contained within the unit circle. DC (0 Hz) or f_{start} is mapped to -1, while f_{end} is mapped to $e^{j\frac{\pi}{4}}$. This is expressed mathematically in (12.8) and graphically in Fig. 12.3.

$$Z_{\text{unit}}(z) = Z\left(\phi \frac{z+1}{z-1}\right), \tag{12.8}$$

where ϕ is chosen such that the following mapping conditions are met:

$$\begin{aligned} \omega &= +1 \quad \rightarrow \theta = +\frac{\pi}{4} \\ \omega &= 0 \quad\;\;\, \rightarrow \theta = \pi \\ \omega &= -1 \quad \rightarrow \theta = -\frac{\pi}{4}. \end{aligned}$$

Furthermore, the projection is performed by computing the Fourier coefficients of $Z_{\alpha\beta}(j\omega)$ on the unit circle through means of a fast Fourier transform (FFT). This is mathematically expressed as

$$Z_{\text{unit}}^{\text{stable}}(\theta) = \sum_{n=0}^{\chi} f_n e^{jn\theta} \qquad Z_{\text{unit}}^{\text{unstable}}(\theta) = \sum_{n=-1}^{-\chi} f_n e^{jn\theta}. \tag{12.9}$$

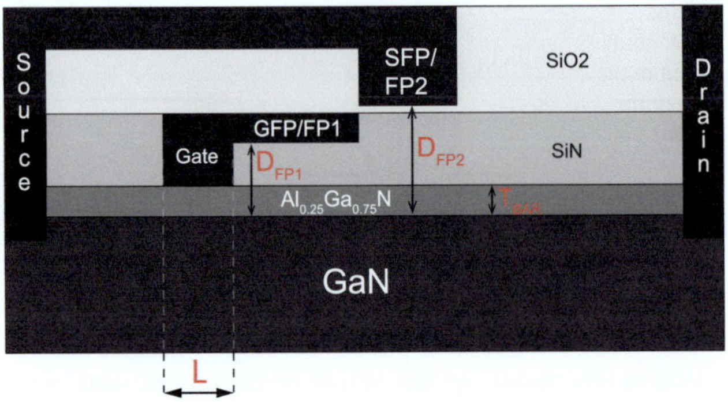

Fig. 12.4 A cross-sectional view of the device that has been modeled. The physical parameters of the model that have been used for the stability analysis have been denoted to aid with the visual representation

Finally, once (12.9) has been calculated, the results are then transformed back to the complex plane where they can be plotted as a traditional pole-zero map for analysis.

12.1.2 Non-Linear Stability Analysis in the Presence of Manufacturing Variations

The methodology to analyze the non-linear stability for the nominal device is extended to account for the manufacturing variations. In Fig. 12.4, a cross-sectional representation of the device being modeled is shown. Also depicted are the device dimensions that are varied statistically to represent the manufacturing variations. In addition to device geometry parameters, key electrical parameters are also statistically varied. The parameters that are statistically varied are: device cut-off voltage V_{off}, barrier layer thickness T_{bar}, low-field mobility u_0, cut-off voltage of field-plate region $V_{offfp1,2}$, and distance of field-plate region to 2-DEG $D_{fp1,2}$. These parameters have been selected as they cause large variation in the overall electrical I-V and C-V characteristics of the device. The methodology presented here is extendable to include more parameters.

12.1.3 Nominal Device Non-Linear Stability Analysis

On the selected geometry and electrical parameters, the effect of manufacturing variations is modeled by allowing statistical variations on the parameters. Model parameters are varied with a Gaussian distribution that has a standard deviation

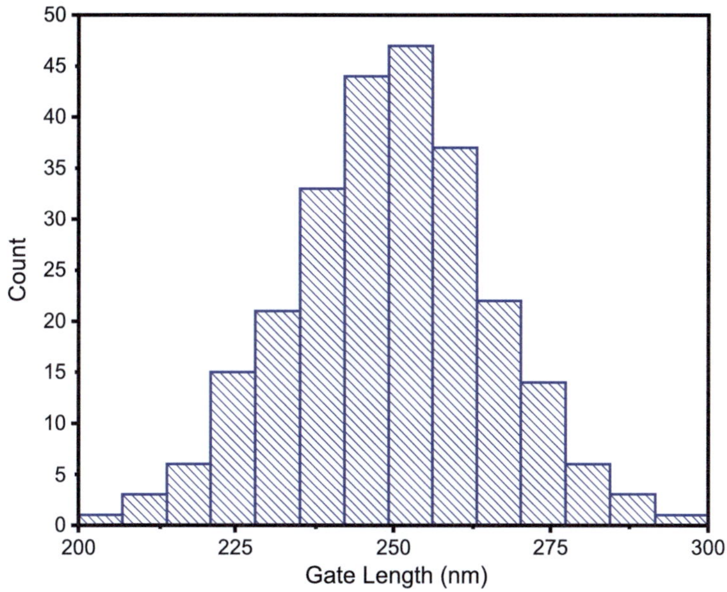

Fig. 12.5 Histogram plot showing the variance of the model parameter L, where μ = 250 and $\sigma\%$ = 20, using a Gaussian distribution

Table 12.1 Physical model parameter list, with description, used for the non-linear stability analysis. Their means (μ) and standard deviation percentage ($\pm\sigma\%$) are also given

Parameter	Description	μ	σ %
L (nm)	Gate length	250	20
T_{BAR} (nm)	AlGaN barrier thickness	25	35
V_{OFF} (V)	Cut-off voltage	−2.78	30
U_0 (cm^2/V.sec)	Low-field mobility	2130	30
DFP1 (nm)	Distance of FP1 from 2-DEG charge	35	35
V_{OFFFP1} (V)	FP1 cut-off voltage	−14.24	30
DFP2 (nm)	Distance of FP2 from 2-DEG charge	851	25
V_{OFFFP2} (V)	FP2 cut-off voltage	−71.02	30

($\sigma\%$) on each parameter. For instance, a graphical representation of the distribution for the gate length is depicted as a histogram plot in Fig. 12.5. We have selected parameters whose variations are statistically significant for this study. For instance, the width (W) of GaN devices often used within PA design is typically quite large—1 mm. Therefore, for a fabrication tolerance of 20%, the variations that would be seen would be less significant for W =1 mm as opposed to L =0.25 um in order to meet the required yield specifications. Thus, parameters such as W for which statistical variations are less significant are not considered in this chapter. Table 12.1 conveys the selected process parameters of the physical model that have been used for the analysis in this chapter (Fig. 12.6).

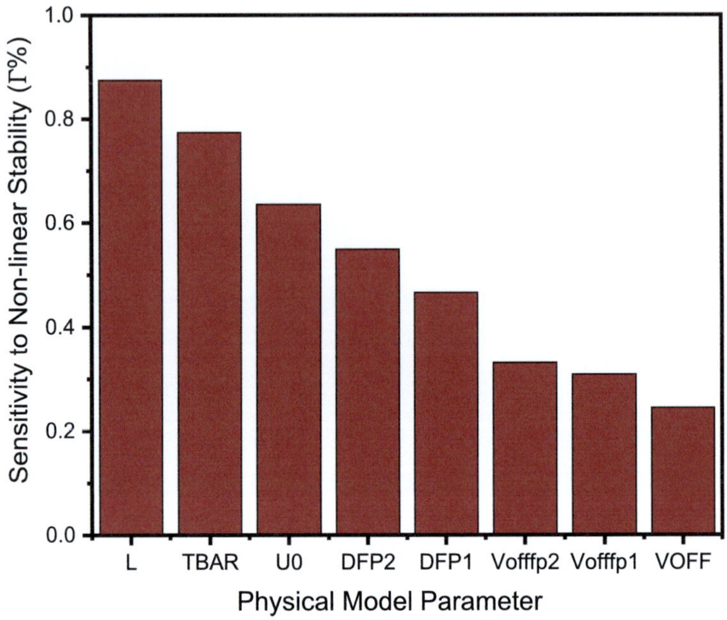

Fig. 12.6 Bar graph representation of (12.10) arranged in the descending order of most sensitive (left) to least sensitive (right)

Non-linear stability analysis in the Ku band (12–18 GHz) with swept input power (P_{IN}) of 0–20 dBm is performed with the developed model. The frequency range used for the analysis was set from 0.1 to 50 GHz to gain further insight as to how the stability of the device shifts when it is pushed beyond its intended limits. As previously mentioned, the nature of the harmonic balance simulations used in this work forms that of a large signal small-signal analysis; therefore, it is essential that a mix equation is incorporated within the simulation in order to ascertain $Z^{n}_{\alpha\beta}(j\omega)$. This can be seen by the "Measurement Equation" (meas1) in Fig. 12.1. Once $Z_{\alpha\beta}(j\omega)$ has been obtained, the Python script is executed in order to perform necessary computations that have been explained above. The results are returned in a tabular format that gives the real and imaginary parts of system's poles and zeros that can then be plotted for the analysis. Figures 12.7, 12.8, 12.9, 12.10, 12.11, 12.12, 12.13, and 12.14 depict the pole-zero maps of the parameter list from Table 12.1. The green symbols with poles on the left-side represent a stable device, while red symbols on the right-side shows instability. It can be seen from Figs. 12.7, 12.8, 12.9, 12.10, 12.11, 12.12, 12.13, and 12.14 that manufacturing process variations move the poles and can cause instabilities.

By analyzing the results from Fig. 12.6, several insights into how the non-linear stability of the system is affected by the tolerances of the device fabrication process can be obtained. The vertical arrows in Figs. 12.7 and 12.8 indicate the deviation of the associated physical model parameter from its mean value with an up arrow

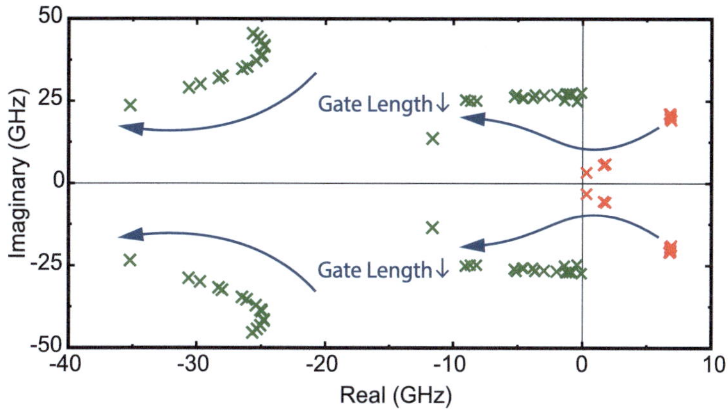

Fig. 12.7 Gate length (L)

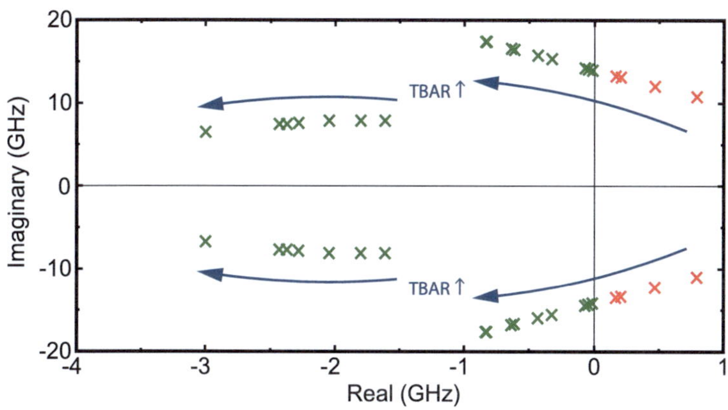

Fig. 12.8 Barrier thickness (TBAR)

(down arrow) indicating an increase (decrease) from the mean value. The horizontal arrows indicate the direction in which the poles are shifted as the parameter changes due to the manufacturing variations. In order to analyze the sensitivity of poles w.r.t. a parameter variation, we define Γ factor as

$$\Gamma = \left| \frac{\Delta \mathrm{X}}{\sigma_n \%} \right|, \tag{12.10}$$

where Γ is the physical model parameters sensitivity for non-linear stability criteria, ΔX is the total shift of the system poles, and $\sigma_n\%$ is the standard deviation percentage for each parameter denoted by n. Γ represents how sensitive device stability is to a particular parameter variation, so a lower Γ represents a parameter that is less crucial for device instability as compared to a parameter with higher Γ.

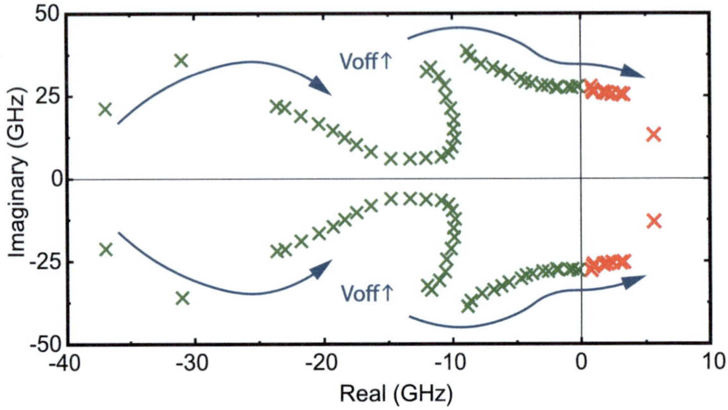

Fig. 12.9 Threshold voltage (Voff)

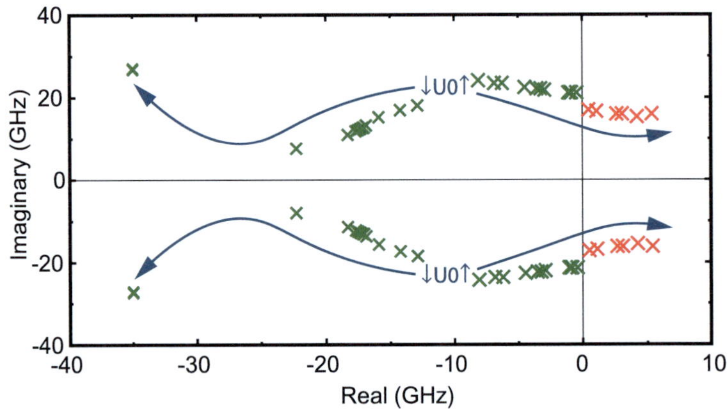

Fig. 12.10 Electron mobility

By utilizing (12.10) for the parameters listed in Table 12.1, insights into how the tolerances of the fabrication process affect the device non-linear stability performance can be drawn. The results of this analysis are depicted in Fig. 12.6, where the parameters have been plotted arranging from most sensitive (left) to least sensitive (right). It is found that the change in channel length is the most sensitive parameter and manufacturing variation on L can change poles of the device significantly to cause instabilities. Other parameters are also sensitive to a lesser degree as shown in Fig. 12.6.

These figures show pole-zero maps for the parameter list given in Table 12.1. The horizontal arrows indicate the movement of the poles with respect to the parameter variation. The vertical arrows indicate the increase/decrease of the associated parameter value. It is to be noted that the corresponding zeros have been omitted

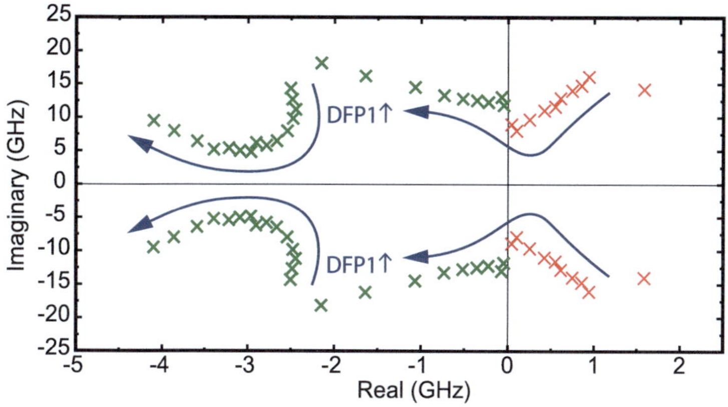

Fig. 12.11 Distance of field-plate 1 to 2-DEG charge

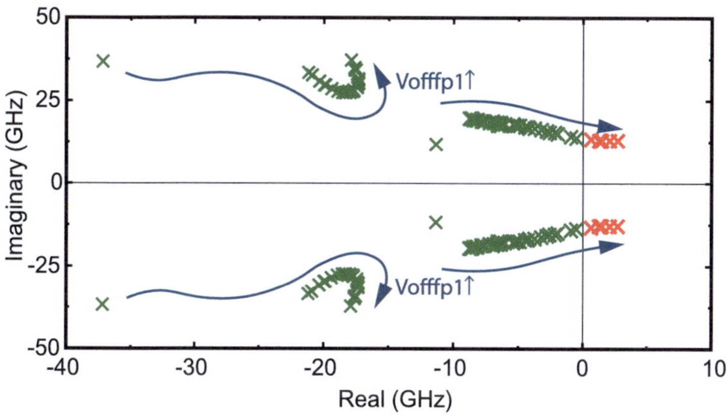

Fig. 12.12 Threshold voltage of field-plate 1

from the plots for simplicity purposes. Furthermore, the frequency range has been selected in order to best illustrate the shift of the poles leading to instability.

12.2 GaN Power Device Circuit Simulations Using ASM-HEMT

This section describes the simulation of a common power electronics circuit, a buck-boost converter with the ASM-HEMT model. Circuit simulations and the ability to use the model features to analyze circuit results are shown. The boost converter circuit is shown in Fig. 12.15. We used specifications of a 100 W boost converter design given and industry standard physics-based compact model for GaN

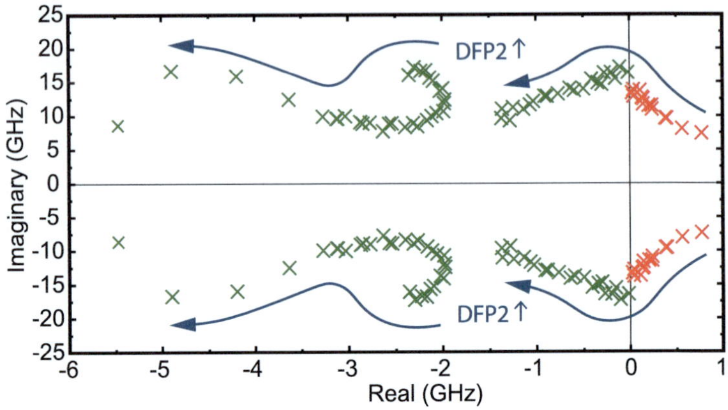

Fig. 12.13 Distance of field-plate 2 to 2-DEG charge

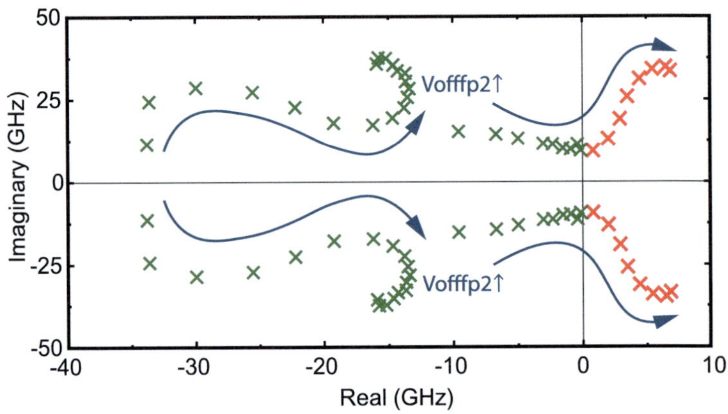

Fig. 12.14 Threshold voltage of field-plate 2

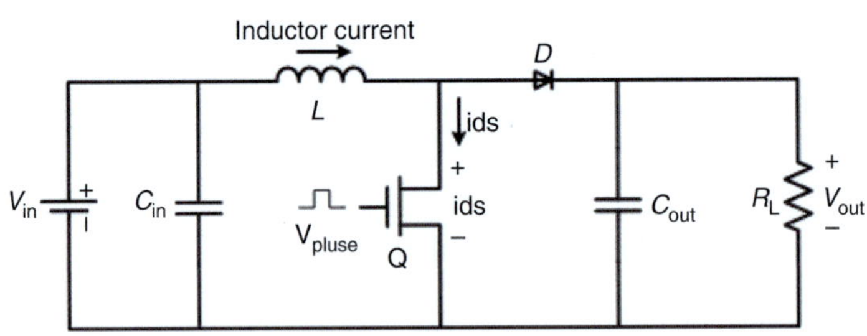

Fig. 12.15 Schematic of boost converter

HEMT: ASM-GaN to simulate the steady state and switching characteristics of such a circuit. For the GaN transistor Q in Fig. 12.15, we used PGA26E19BA, which is a 600 V enhancement-mode (E-mode) GaN MOSFET from Panasonic.

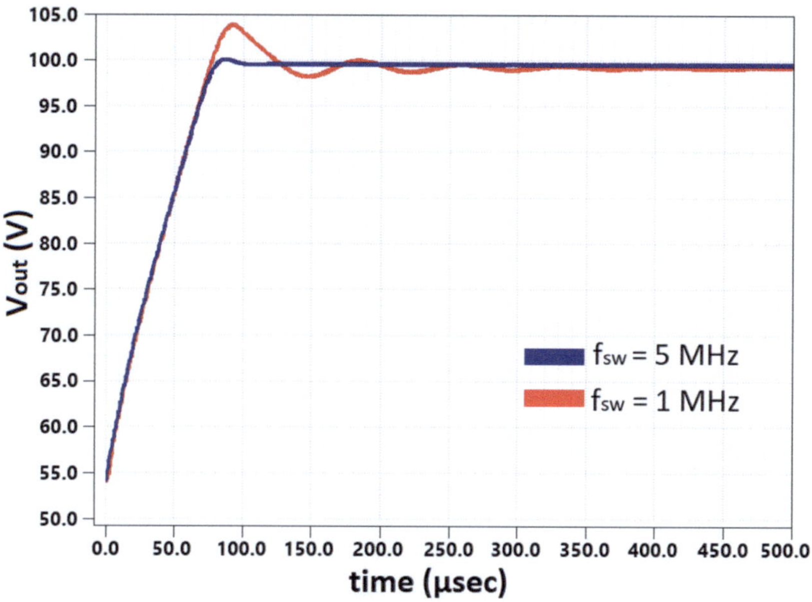

Fig. 12.16 Transient simulation of boost converter

The model parameters were first extracted from the OFF- state C-V and output I-V characterization data. The resultant model card was used in circuit simulation. The circuit was designed for two switching frequencies f_{sw} of 1 MHz and 5 MHz using the standard boost converter equations. The designed boost converter circuit was simulated in a circuit simulator. A transient analysis was run starting from zero time till 1 ms to ensure a steady-state operation of the circuit. Figure 12.16 shows a plot of V_{out} vs. time obtained from the transient simulation. The circuit predicts faster settling time and close to a critically damped response for higher fsw (5 MHz), while a slow settling time and underdamped response for lower fsw (1 MHz).

Figures 12.17 and 12.18 show a zoomed in portion of the steady-state waveform for drain–source voltage (vds) and current (ids) and their product that is the power loss in transistor Q. The switching losses (both ON and OFF) can be obtained from these waveforms. Also, the impact of different field-plates on switching losses can be analyzed by simulating the model with field-plate switches turned off and on.

In order to understand how the OFF-state output capacitor Coss affects the performance of the circuit, we consider the following scenario. During the OFF state of transistor, the given boost topology looks like Fig. 12.19 with an output capacitance C_{oss} value of approximately 67 pF. During this time, the Coss charges with the polarity as shown in Fig. 12.19 with a charging current of $i_{ds,OFF}$, while the output capacitor C_{out} and load resistor R_L pull out currents i_c and i_{load}, respectively. This charged capacitance does not allow $i_{ds,OFF}$ to change suddenly when the transistor Q is switched from OFF state to ON state because of the polarity of its

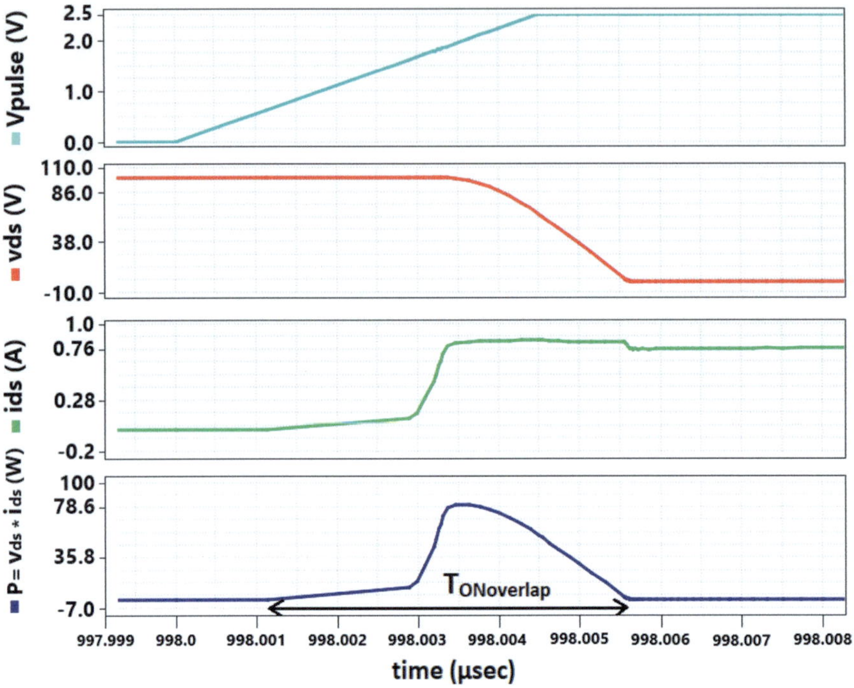

Fig. 12.17 Transient simulation of boost converter

terminals, and hence, the current does not rise from $i_{ds,OFF}$ during the initial ramp up time as shown in Fig. 12.19.

A finite amount of power is lost during switching processes that occur from OFF to ON and then ON to OFF states, with a corresponding energy loss of EON and EOFF, respectively. These energy losses can be quantitatively calculated by integrating the overlap of drain current and drain–source voltage as shown in Figs. 12.17 and 12.18.

$$E_{ON} = \int_{O} FF^{O} Ni(t) \cdot v(t) \cdot dt \tag{12.11}$$

$$E_{OFF} = \int_{O} N^{O} FFi(t) \cdot v(t) \cdot dt. \tag{12.12}$$

It is found that there is no difference between E_{OFF}'s predicted by using model with or without field-plate, while E_{ON} shows a difference of about 8.9 nJ. This difference arises because C_{oss} sees a different capacitance value during the transient voltage drop from 100 V to linear region, which leads to different amount of switching drain currents $i(t)$ during the transition from OFF state to ON state. These varying amounts of switching currents lead to different instantaneous values

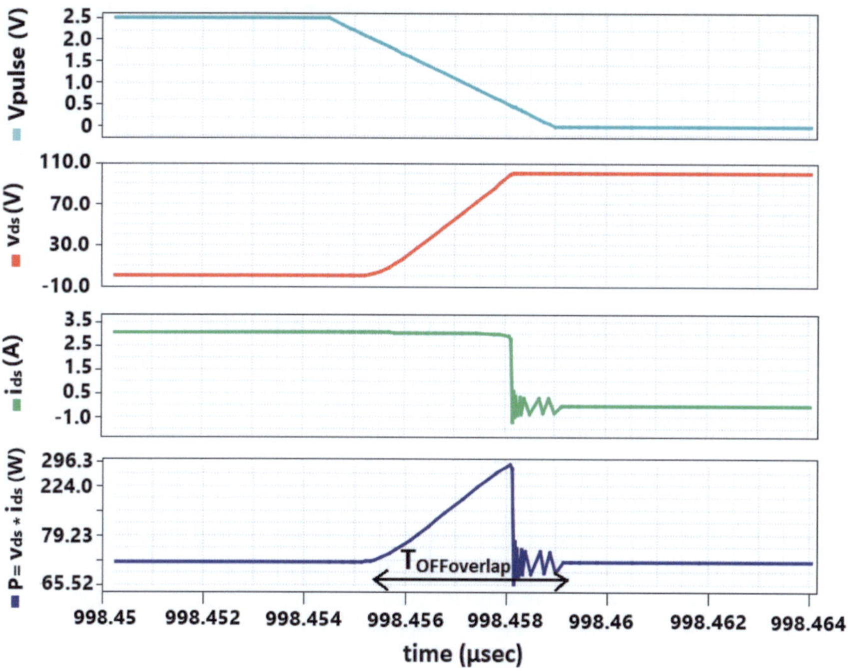

Fig. 12.18 Transient simulation of boost converter

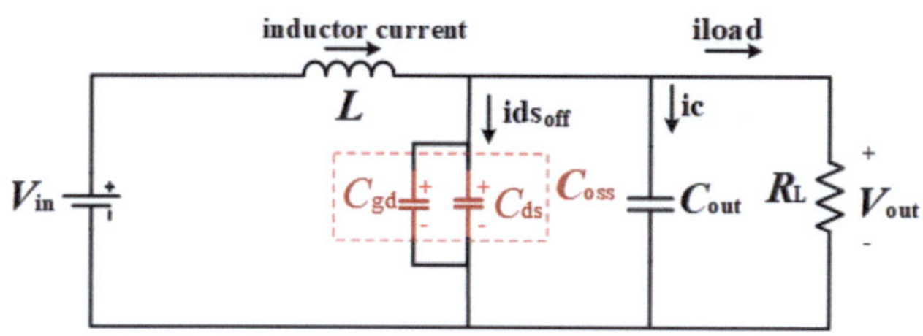

Fig. 12.19 Circuit schematic of boost converter

of switching loss and hence different values of E_{ON}. Thus we can predict the dynamic behavior of a given power circuit using the ASM-GaN model and predict switching losses for a given device. The model allows us to see the difference in losses with and without field-plates. This methodology can be further refined and used to predict switching performance of other topologies, which could be a topic of future work.

12.3 Design Technology Co-optimization Using ASM-HEMT

The physical basis of the ASM-HEMT model can be used for exploring the effect of device features in circuit level performance metrics. This offers new possibilities to optimize device and design simultaneously. The design space is expanded with this capability. The sub-sections below discuss two specific examples where the non-linear AM/PM is connected to the device feature.

12.3.1 Barrier Thickness Dependence of AM/PM

To study the dependence of GaN HEMT non-linear behavior on T_{bar}, we model five GaN HEMTs fabricated with same fabrication conditions and varying T_{bar}. These devices have a source–drain L_{sd} spacing of 3 μm and a gate length L_g of 1 μm and device width $W = 50\,\mu\text{m}$. We model all these devices using a physics-based compact model. The geometrical device parameters like L_g, T_{bar}, L_{sd}, and W values are directly fed into the model as these are known from device fabrication details. Drain current model contains physically grounded parameters that are extracted for all the five GaN HEMT devices using the extraction procedure presented in the last chapter.

The model results are shown in Fig. 12.20 for the drain current I_d and its first derivative trans-conductance G_m, second derivative G'_m, and third derivative G''_m for all five devices. For understanding non-linear behavior of the device, it is critical not only to model I_d and G_m but also the higher order derivatives of the drain current. As shown in Fig. 12.1, accurate modeling of second- G'_m and third-order G''_m derivatives using the surface-potential based model has been obtained. The higher order derivatives are used to analyze the non-linear behavior and the observed trends help developing insights into how non-linearities change as the device as T_{bar} is varied. It is found that the extracted model parameters vary systematically with T_{bar} as shown in Fig. 12.21. The variation in the cut-off voltage V_{off} with T_{bar} follows similar trends as expected from the charge control physics. V_{off} is the model parameter representing strength of the 2-DEG charge density for applied V_g. Lower V_{off} indicates higher 2-DEG charge density at fix V_g and vice versa. We see a non-linear trend in V_{off} with T_{bar} (see Fig. 12.2), with V_{off} saturating as T_{bar} approaches 25 nm. This has been studied previously and was attributed to the fact that 2-DEG charge density starts approaching polarization charge density. After which, 2-DEG charge density does not increase significantly even if T_{bar} is increased. This systematic trend captured by the model highlights its usefulness as a tool to connect device technology to the circuit design, as we can use this model to see the impact of the device trends at the circuit level. More interestingly, we also see variations in effective mobility parameters vary with T_{bar} in our extractions as shown in Fig. 12.21. Effective mobility accounts for the carrier scattering mechanisms in the 2-DEG channel and is typically lower than the low-

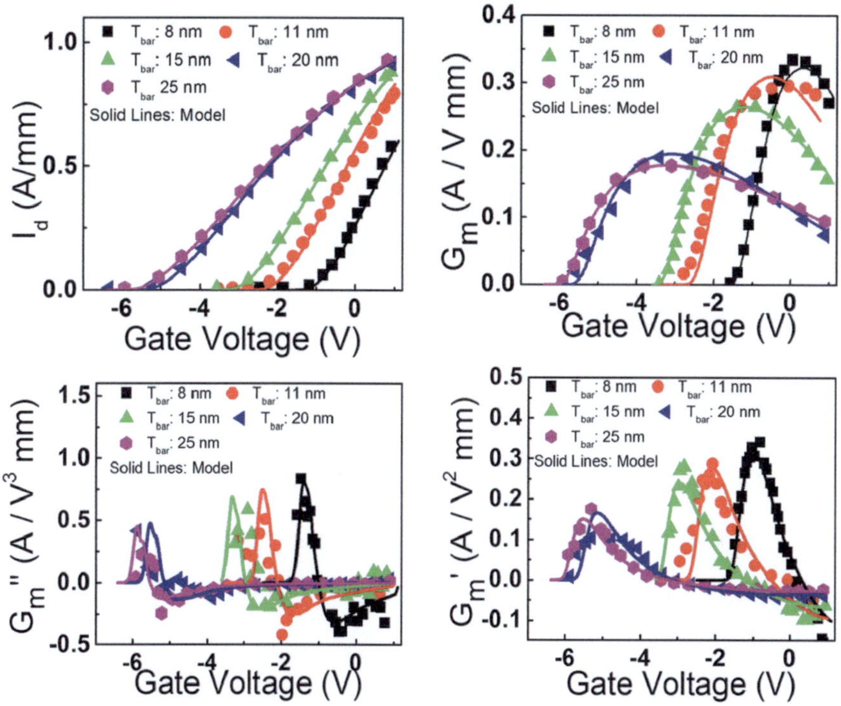

Fig. 12.20 Excellent model agreement for drain current I_d, its first derivative trans-conductance G_m, second derivative G_m', and the third derivative G_m'' obtained for five GaN HEMTs with varying barrier layer thicknesses from 8 nm to 25 nm. Symbols: Measured data and solid lines: Model

field mobility that is measured by Hall measurements. After accurately modeling devices with five different barrier later thicknesses as shown in Fig. 12.8, we see trends in low-field mobility $U0$ with T_{bar}. $U0$ increases with T_{bar} till $T_{bar} = 15$ nm and then due to barrier relaxation and dislocation formation $U0$ drops. This window for high mobility in GaN HEMTs has been identified in other works too. The model captures this physical trend quite well. From non-linearity point of view, the variation in parameter UB is quite interesting. While UA remains constant, UB is found to increase with a decrease in Tbar. An increase in UB indicates a stronger dependence of effective mobility on the input signal amplitude. This in turn translates into larger non-linearity. This is presented in Fig. 12.3 that shows the variation in OIP3 as a function of barrier layer thickness for these five devices as obtained from harmonic balance simulations. The simulations are performed using the accurately fitted model for class-A bias condition for same DC bias current in all the devices. The bias-independent parasitic components are assumed to typical values as obtained in [6]. Note that the parasitic components do not significantly affect the non-linearity behavior trends seen in these simulations as they are bias-independent. Figure 12.4 shows that OIP3 systematically decreases with a decrease

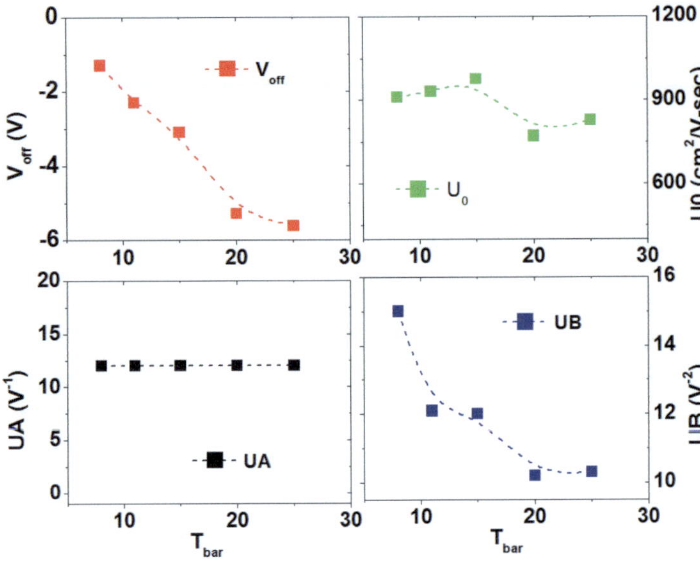

Fig. 12.21 Systematic variation of model parameters Voff, U0, UA, and UB with barrier layer thickness as obtained after modeling five different GaN HEMT devices with Tbar varying from 8 nm to 25 nm

Fig. 12.22 Variation of the OIP3 with barrier layer thickness for five different devices. As barrier thickness reduces, device becomes increasingly non-linear

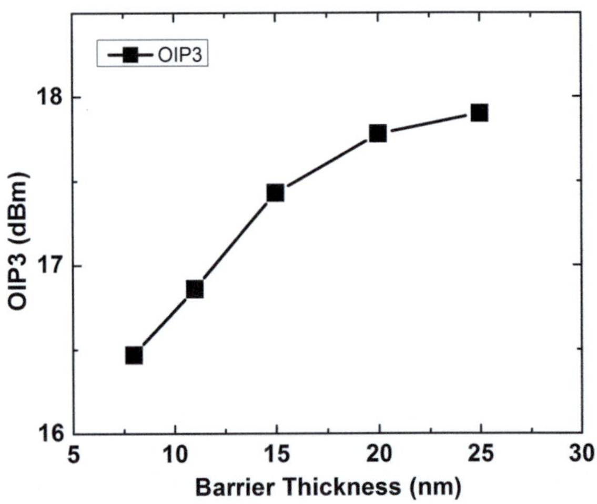

in Tbar. This is due to increasing UB indicating its stronger non-linear behavior as Tbar decreases. A thin Tbar causes effective mobility to depend strongly on input amplitude. On the other hand, a thin Tbar helps to get a higher Gm, better gain, and output power. This chapter reveals this trade-off for the device technology developer (Fig. 12.22).

Another important non-linearity is the phase non-linearity. Non-linear behavior of the phase of the output signal depends strongly on the variation of gate-to-source C_{gs} and gate-to-drain C_{gd} capacitance as a function of the input amplitude. C_{gd}

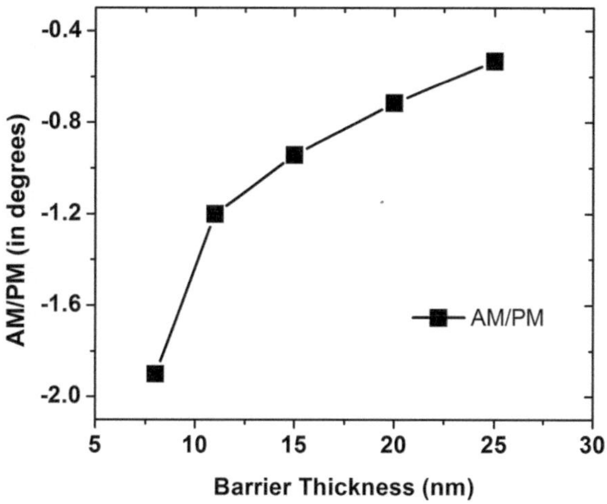

Fig. 12.23 Variation of AM/PM non-linearity increases with decrease in barrier layer thickness due to the larger change in capacitances with input signal

and C_{gd} vary with gate voltage. Minimum values of these capacitances $C_{gs,min}$ and $C_{gd,min}$ are observed for gate voltage lower than cut-off voltage. These capacitances increase rapidly with V_g after cut-off voltage, and they reach their maximum values $C_{gs,max}$ and $C_{gd,max}$ for V_g beyond cut-off voltage when there is a strong 2-DEG formation. While the minimum values $C_{gs,min}$ and $C_{gd,min}$ are primarily decided by parasitic/fringe capacitance, the maximum values depend strongly on T_{bar}. A thin T_{bar} produces higher maximum value of these capacitances as compared to a thicker one. This behavior of capacitance is captured by the model, owing to its physical nature. It can be seen from Figs. 12.4 and 12.5 that maximum values of Cgs and Cgd increase with a decrease in T_{bar} indicating that there will be a larger variation in capacitance $C_{gs,max} - C_{gs,min}$ possible for a thinner T_{bar} device. This directly affects the phase non-linearity or AM/PM behavior. In Fig. 12.23, we show simulation of AM/PM behavior at a fixed input power level of 20 dBm, biased at same drain current for all the five GaN devices we have modeled. It is clear that AM/PM non-linearity degrades as Tbar is decreased for the aforementioned reason. This reveals another important trade-off to the device developer. A decrease in Tbar causes larger AM/PM non-linearity; however, it also provides a larger Gm. It is very interesting to see that due to the physical basis of the model it can be used as a tool to identify these trade-offs during device development. This provides an excellent means for device engineer to gain insights into manifestation of device behavior at the circuit level. The model can serve as a tool to tune and optimize device to circuit level performance.

12.3.2 Field-Plate Dependence of AM/PM

For devices with field-plates, the charge modulation in the field-plate region can also lead to non-linearity under large signal operation. This non-trivial connection

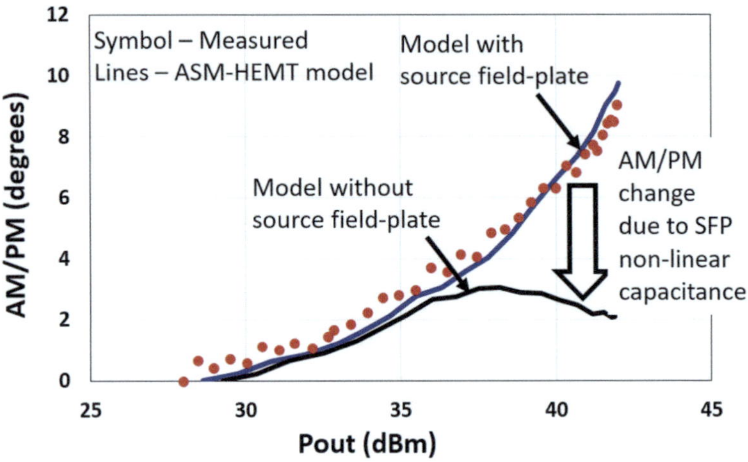

Fig. 12.24 Measured and modeled AM/PM versus output power Pout. Model simulations with constant capacitor and without source field-plate show that impact of source field-plate non-linear capacitance on AM/PM non-linearity

between device feature and the circuit parameter is enabled with the physics-based formulations. To understand the effect of field-plates, first a large signal model is developed. Once an accurate model for the non-linearity AM/PM has been obtained, the models for the field-plate within the ASM-HEMT model can be turned off to see their impact. In Fig. 12.24, good model agreement between the measured and modeled AM/PM behavior can be seen. This is achieved with the accurate bias-dependent capacitor model. To analyze the impact of source field-plate on AM/PM behavior, model simulations are performed with source field-plate turned off. ASM-HEMT model has a model switch FPMOD, and setting this to zero removed the source field-plate (SFP) region from the device. Removing field-plate naturally reduced the output capacitance Cds and disturbed the model results from drain–source capacitance C_{ds}. To keep C_{ds} model fitting to be the same, a constant, i.e., bias-independent capacitance, is added in the model. AM/PM simulations are performed with this model that does not have source field-plate, nevertheless gives the same small-signal model results. It is clear from Fig. 12.24 that removal of SFP has a significant impact on AM/PM at large output power.

This can be explained with large signal swings of the drain–source voltage Vds shown in Fig. 12.25. Vds goes beyond the cut-off voltage of source field-plate region $V_{off,SFP} \approx -80$ V, as a result field-plate region charge gets modulated significantly during the large signal swing. This charge modulation causes a large variation in field-plate capacitance leading to a higher non-linearity in the output capacitance. To confirm this, we plot the phase out at the output terminals versus power in Fig. 12.26. A larger change in output phase with output power can be seen for model with SFP as compared to model without SFP from Fig. 12.26. Although model without SFP has larger phase, it is relatively constant as compared to the model with SFP. SFP

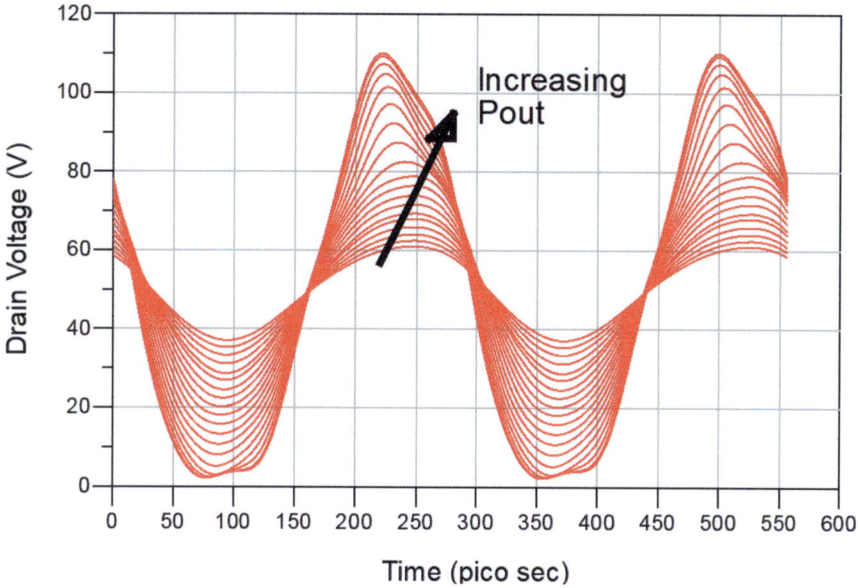

Fig. 12.25 Signal swing in V_{ds} with output power. Large V_{ds} swings cause non-linear change in source field-plate capacitance and result in AM/PM non-linearity shown in Fig. 12.24

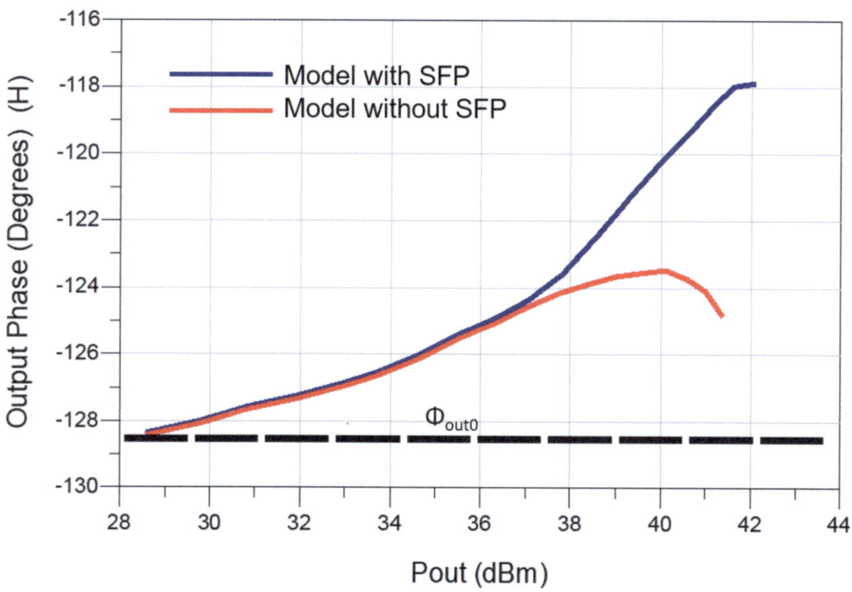

Fig. 12.26 Change in output phase $\Delta\Phi_{out}$ with P_{out} is larger in the presence of source field-plate due to SFP adding non-linear capacitance. Without SFP phase is larger in magnitude, but it does not change as much as in the presence of SFP

causing non-linear output capacitance is the source of this phase variation and hence larger AM/PM. Note that input phase Φ_{in} also changes with Pout; however, Φ_{out} is found to be much larger. ASM-HEMT model with physical formulation is shown to serve as a very useful tool to gain deeper insights into non-linear device behavior. This can enable device optimization for meeting non-linearity specifications.

Chapter 13
Resources for ASM-HEMT Model Users

13.1 GITHUB Repository

The model code, documentation, and other useful resources on this model are available in one GITHUB repository. If you are new to GIT, see https://git-scm.com/about/. If you are new to GITHUB, see https://github.com/.

The GITHUB repository is the central resource that is regularly updated and managed by the author. At the time of writing, this repository has the following resources for the users. Note that the repository is regularly updated, and at the time of reading, there might be more information on the repository than what is described below:

- Model code: The model code in Verilog-AMS programming language is available in the directory **vacode** within the repository. The version of the model code available is the latest open source industry standard version of the ASM-HEMT model. As the model is further developed, new versions will be released and will be added to this directory.
- Technical manual: Technical manual of the model in the form of a PDF file is available in the repository. Technical manual is a concise documentation of the model.
- Videos: Over time several talks and presentations on this model have been created. Some of these are also in the form of videos uploaded to YouTube. This folder contains all the links to the videos of this model.
- Research papers: Active research is on-going on this model. The past and latest research papers on this model are added to folder **research papers** of the repository.

S. Khandelwal, *Advanced SPICE Model for GaN HEMTs (ASM-HEMT)*,
https://doi.org/10.1007/978-3-030-77730-2_13

References

1. Nakamura, S.: Jpn. J. Appl. Phys. **30**(10A), L1705 (1991)
2. Nakamura, S., Senoh, M., Mukai, T.: Jpn. J. Appl. Phys. **32**(1A), L8 (1993)
3. Nakamura, S., Mukai, T., Senoh, M.: Jpn. J. Appl. Phys. **30**(12A), L1998 (1991)
4. Khan, M.A., Hu, X., Sumin, G., Lunev, A., Yang, J., Gaska, R., Shur, M.: IEEE Electron Device Lett. **21**(2), 63 (2000)
5. Khan, M.A., Van Hove, J., Kuznia, J., Olson, D.: Appl. Phys. Lett. **58**(21), 2408 (1991)
6. Amano, H., Sawaki, N., Akasaki, I., Toyoda, Y.: Appl. Phys. Lett. **48**(5), 353 (1986)
7. Hiramatsu, K., Itoh, S., Amano, H., Akasaki, I., Kuwano, N., Shiraishi, T., Oki, K.: J. Cryst. Growth **115**(1–4), 628 (1991)
8. Lanford, W., Tanaka, T., Otoki, Y., Adesida, I.: Electron. Lett. **41**(7), 449 (2005)
9. Binari, S., Kruppa, W., Dietrich, H., Kelner, G., Wickenden, A., Freitas Jr, J.: Solid-State Electron. **41**(10), 1549 (1997)
10. Zhang, N.Q., Keller, S., Parish, G., Heikman, S., DenBaars, S., Mishra, U.: IEEE Electron Device Lett. **21**(9), 421 (2000)
11. Saito, W., Takada, Y., Kuraguchi, M., Tsuda, K., Omura, I.: IEEE Trans. Electron Devices **53**(2), 356 (2006)
12. Palacios, T., Dora, Y., Chakraborty, A., Sanabria, C., Keller, S., DenBaars, S., Mishra, U.: Optimization of AlGaN/GaN HEMTs for High Frequency Operation (2006)
13. Micovic, M., Brown, D., Regan, D., Wong, J., Tang, Y., Herrault, F., Santos, D., Burnham, S., Tai, J., Prophet, E. et al.: In: *Proceedings of the 2016 IEEE International Electron Devices Meeting (IEDM)*, pp. 3–3. IEEE, New York (2016)
14. Wu, Y.F., Saxler, A., Moore, M., Smith, R., Sheppard, S., Chavarkar, P., Wisleder, T., Mishra, U., Parikh, P.: IEEE Electron Device Lett. **25**(3), 117 (2004)
15. Wu, Y.F., Moore, M., Wisleder, T., Chavarkar, P., Mishra, U., Parikh, P.: In: *IEDM Technical Digest: IEEE International Electron Devices Meeting, 2004*, pp. 1078–1079. IEEE, New York (2004)
16. Martín-Guerrero, T.M., Santarelli, A., Gibiino, G.P., Traverso, P.A., Camacho-Peñalosa, C., Filicori, F.: IEEE Trans. Microwave Theory Tech. **68**(5), 1627 (2020)
17. Raffo, A., Bosi, G., Vadalà, V., Vannini, G.: IEEE Trans. Microwave Theory Tech. **62**(1), 73 (2013)
18. Leckey, J.: In *Proceedings of the 2011 6th European Microwave Integrated Circuit Conference*, pp. 13–16. IEEE, New York (2011)
19. Xu, J., Jones, R., Harris, S.A., Nielsen, T., Root, D.E.: In: *Proceedings of the 2014 IEEE MTT-S International Microwave Symposium (IMS2014)*, pp. 1–3, IEEE, New York (2014)
20. Sadi, T., Schwierz, F.: Solid-State Electron. **81**, 91 (2013)

S. Khandelwal, *Advanced SPICE Model for GaN HEMTs (ASM-HEMT)*,
https://doi.org/10.1007/978-3-030-77730-2

21. Melczarsky, I., Lonac, J.A., Filicori, F., Santarelli, A.: IEEE Trans. Microwave Theory Tech. **56**(9), 2017 (2008). https://doi.org/10.1109/TMTT.2008.2001956
22. Wang, Q., Wu, Y., Cong, S., Yu, Y., Zhao, C., Liu, H., Tang, H., Xu, Y., Kang, K.: In: *Proceedings of the 2018 Asia-Pacific Microwave Conference (APMC)*, pp. 348–350 (2018). https://doi.org/10.23919/APMC.2018.8617163
23. Schroter, M., Haferlach, M., Sakalas, P., Claus, M.: In: *2013 IEEE MTT-S International Microwave Symposium Digest (MTT)*, pp. 1–4 (2013). https://doi.org/10.1109/MWSYM.2013.6697679
24. Ji, D., Li, W., Chowdhury, S.: In *Proceedings of the 2018 IEEE 30th International Symposium on Power Semiconductor Devices and ICs (ISPSD)* (2018), pp. 208–211. https://doi.org/10.1109/ISPSD.2018.8393639
25. Trojman, L., Vaca, A., Procel, L.M.: In: *Proceedings of the 2018 IEEE XXV International Conference on Electronics, Electrical Engineering and Computing (INTERCON)*, pp. 1–4 (2018). https://doi.org/10.1109/INTERCON.2018.8526445
26. Aguirre, P., Rau, M., Schenk, A.: In: *Proceedings of the 2018 Joint International EUROSOI Workshop and International Conference on Ultimate Integration on Silicon (EUROSOI-ULIS)*, pp. 1–4 (2018). https://doi.org/10.1109/ULIS.2018.8354744
27. Marinković, Z., Crupi, G., Caddemi, A., Marković, V.: In: *Proceedings of the 10th Symposium on Neural Network Applications in Electrical Engineering*, pp. 89–92 (2010). https://doi.org/10.1109/NEUREL.2010.5644101
28. Xu, J., Gunyan, D., Iwamoto, M., Cognata, A. Root, D.E.: In: *Proceedings of the 2006 IEEE MTT-S International Microwave Symposium Digest*, pp. 469–472 (2006). https://doi.org/10.1109/MWSYM.2006.249594
29. Xu, J., Gunyan, D., Iwamoto, M., Horn, J.M., Cognata, A., Root, D.E.: In: *Proceedings of the 2007 IEEE/MTT-S International Microwave Symposium*, pp. 2011–2014 (2007). https://doi.org/10.1109/MWSYM.2007.380244
30. Xu, J., Root, D.E.: In: *Proceedings of the 2015 IEEE MTT-S International Conference on Numerical Electromagnetic and Multiphysics Modeling and Optimization (NEMO)*, pp. 1–3 (2015). https://doi.org/10.1109/NEMO.2015.7415102
31. Khandelwal, S.: In: *Proceedings of the 2019 IEEE 20th Wireless and Microwave Technology Conference (WAMICON)*, pp. 1–4 (2019). https://doi.org/10.1109/WAMICON.2019.8765472
32. Khandelwal, S., Fjeldly, T.A.: In: *Proceedings of the 2012 24th International Symposium on Power Semiconductor Devices and ICs*, pp. 241–244 (2012). https://doi.org/10.1109/ISPSD.2012.6229068
33. Rakheja, S., Antoniadis, D.: In: *Proceedings of the 2015 IEEE International Electron Devices Meeting (IEDM)*, pp. 28.6.1–28.6.4 (2015). https://doi.org/10.1109/IEDM.2015.7409790
34. Khandelwal, S., Chauhan, Y.S., Lu, D.D., Venugopalan, S., Ahosan Ul Karim, M., Sachid, A.B., Nguyen, B.Y., Rozeau, O., Faynot, O., Niknejad, A.M., Hu, C.C.: IEEE Trans. Electron Devices **59**(8), 2019 (2012). https://doi.org/10.1109/TED.2012.2198065
35. Gildenblat, G., Li, X., Wu, W., Wang, H., Jha, A., Van Langevelde, R., Smit, G.D.J., Scholten, A.J., Klaassen, D.B.M.: IEEE Trans. Electron Devices **53**(9), 1979 (2006). https://doi.org/10.1109/TED.2005.881006
36. Bucher, M., Lallement, C., Enz, C., Krummenacher, F.: In: *Proceedings of the 1996 IEEE International Symposium on Circuits and Systems. Circuits and Systems Connecting the World. ISCAS 96*, vol. 4, pp. 703–706 (1996). https://doi.org/10.1109/ISCAS.1996.542121
37. Horn, J., Root, D.E., Simpson, G.: In: *Proceedings of the 2010 IEEE Compound Semiconductor Integrated Circuit Symposium (CSICS)*, pp. 1–4. IEEE, New York (2010)
38. Khandelwal, S., Goyal, N., Fjeldly, T.A.: IEEE Trans. Electron Devices **58**(10), 3622 (2011)
39. Khandelwal, S., Chauhan, Y.S., Fjeldly, T.A.: IEEE Trans. Electron Devices **59**(10), 2856 (2012)
40. Albahrani, S.A., Mahajan, D., Hodges, J., Chauhan, Y.S., Khandelwal, S.: IEEE Trans. Electron Devices **66**(1), 87 (2018)
41. Yin, H., Bilbro, G., Trew, R.: In: *Proceedings of the 2007 IEEE/MTT-S International Microwave Symposium*, pp. 787–790. IEEE, New York (2007)

42. Martin, P., Lucci, L.: In: *Proceedings of the 20th International Conference Mixed Design of Integrated Circuits and Systems-MIXDES 2013*, pp. 92–95. IEEE, New York (2013)
43. Radhakrishna, U., Imada, T., Palacios, T., Antoniadis, D.: Phys. Status Solidi C **11**(3–4), 848 (2014)
44. Mertens, S.: In: *MOS AK Workshop* (2015)
45. Ibbetson, J.P., Fini, P., Ness, K., DenBaars, S., Speck, J., Mishra, U.: Appl. Phys. Lett. **77**(2), 250 (2000)
46. Kola, S., Golio, J.M., Maracas, G.N.: IEEE Electron Device Lett. **9**(3), 136 (1988)
47. Khandelwal, S., Ghosh, S., Chauhan, Y., Iniguez, B., Fjeldly, T.: In: *Proceedings of the 2015 IEEE Compound Semiconductor Integrated Circuit Symposium (CSICS)*, pp. 1–4. IEEE, New York (2015)
48. Khandelwal, S., Chauhan, Y.S., Fjeldly, T.A.: IEEE Trans. Electron Devices **59**(10), 2856 (2012). https://doi.org/10.1109/TED.2012.2209654
49. Ward, D.E.: (1982)
50. Joh, J., Del Alamo, J.A.: IEEE Trans. Electron Devices **58**(1), 132 (2010)
51. Jin, D., del Alamo, J.A.: In: *Proceedings of the 2012 24th International Symposium on Power Semiconductor Devices and ICs*, pp. 333–336. IEEE, New York (2012)
52. Couvidat, J., Subramani, N.K., Gillet, V., Laurent, S., Charbonniaud, C., Nallatamby, J.C., Prigent, M., Deltimple, N., Quéré, R.: In: *Proceedings of the 2018 IEEE/MTT-S International Microwave Symposium-IMS*, pp. 720–723. IEEE, New York (2018)
53. Bouya, M., Malbert, N., Labat, N., Carisetti, D., Perdu, P., Clément, J.C., Lambert, B., Bonnet, M., Microelectron. Reliab. **48**(8–9), 1366 (2008)
54. Bisi, D., Meneghini, M., Van Hove, M., Marcon, D., Stoffels, S., Wu, T.L., Decoutere, S., Meneghesso, G., Zanoni, E.: Phys. Status Solidi A **212**(5), 1122 (2015)
55. Rathmell, J.G., Parker, A.E.: In: *Microelectronics: Design, Technology, and Packaging III, Proceedings of SPIE Conference on Microelectronics, MEMS and Nanotechnology*, vol. 6798, pp. 67980R(1–11). Canberra, Australia (2007). https://doi.org/10.1117/12.758711
56. Albahrani, S.A., Parker, A.E., Heimlich, M.: In: Solid-State Electron. **126**, 143–151 (2016)
57. Khandelwal, S., Chauhan, Y.S., Fjeldly, T.A., Ghosh, S., Pampori, A., Mahajan, D., Dangi, R., Ahsan, S.A.: Manuscript submitted to IEEE Trans. Electron Devices
58. Shockley, W., Read, W.T., Phys. Rev. **87**(5), 835 (1952). https://doi.org/10.1103/PhysRev.87.835
59. Sah, C.: Proc. IEEE **55**(5), 654 (1967). https://doi.org/10.1109/PROC.1967.5630
60. Jeon, K., Kwon, Y., Hong, S.: IEEE Microwave Guided Wave Lett. **7**(3), 78 (1997). https://doi.org/10.1109/75.556038
61. Yang, P., Epler, B.D., Chatterjee, P.K.: IEEE J. Solid-State Circuits **18**(1), 128 (1983)
62. Chung, S.S.: IEEE Trans. Comput. Aided Des. Integr. Circuits Syst. **8**(1), 1 (1989)
63. Winton, R.S., Bandy, W.R., IEEE Trans. Comput. Aided Des. Integr. Circuits Syst. **17**(7), 631 (1998)
64. Gharabagi, R., El Nokali, M.: Solid-State Electron. **32**(1), 57 (1989)
65. Si, J., Wei, J., Chen, W., Zhang, B.: IEEE Trans. Electron Devices **60**(10), 3223 (2013)
66. Khandelwal, S., Chauhan, Y.S., Fjeldly, T.A.: IEEE Trans. Electron Devices **59**(10), 2856 (2012). https://doi.org/10.1109/TED.2012.2209654
67. Hodges, J., Schwantuschke, D., Raay, F. v., Brückner, P., Quay, R., Khandelwal, S.: In: *Proceedings of the 2019 IEEE MTT-S International Microwave Symposium (IMS)*, pp. 603–606 (2019). https://doi.org/10.1109/MWSYM.2019.8700738
68. Karamarkovic, J., Pesic, T.V., Jankovic, N.: In: *Proceedings of the 2000 International Semiconductor Conference. 23rd Edition. CAS 2000 Proceedings (Cat. No. 00TH8486)*, vol. 1, pp. 311–314. IEEE, New York (2000)
69. Lee, J., Kim, B., Kim, Y., Park, S.: Solid-State Electron. **37**(8), 1485 (1994)
70. Bardés, D., Alcubilla, R.: Solid-State Electron. **41**(8), 1177 (1997)
71. Turuvekere, S., Rawal, D.S., DasGupta, A., DasGupta, N.: IEEE Trans. Electron Devices **61**(12), 4291 (2014). https://doi.org/10.1109/TED.2014.2361436

72. Zhang, H., Miller, E.J., Yu, E.T.: J. Appl. Phys. **99**(2), 023703 (2006). https://doi.org/10.1063/1.2159547
73. Yan, D., Lu, H., Cao, D., Chen, D., Zhang, R., Zheng, Y.: Appl. Phys. Lett. **97**(15), 153503 (2010). https://doi.org/10.1063/1.3499364
74. Turuvekere, S., Karumuri, N., Rahman, A.A., Bhattacharya, A., DasGupta, A., DasGupta, N.: IEEE Trans. Electron Devices **60**(10), 3157 (2013). https://doi.org/10.1109/TED.2013.2272700
75. Ghosh, S., Dasgupta, A., Khandelwal, S., Agnihotri, S., Chauhan, Y.S.: IEEE Trans. Electron Devices **62**(2), 443 (2015). https://doi.org/10.1109/TED.2014.2360420
76. Sze, S.M.: *Physics of Semiconductor Devices*, pp. 402–407. Wiley, New York (2001), chap. MIS diode and charge-coupled devices
77. Maeda, T., Okada, M., Ueno, M., Yamamoto, Y., Kimoto, T., Horita, M., Suda, J.: Appl. Phys. Express **10**(5), 051002 (2017). https://doi.org/10.7567/apex.10.051002
78. Hsu, J.W.P., Manfra, M.J., Lang, D.V., Richter, S., Chu, S.N.G., Sergent, A.M., Kleiman, R.N., Pfeiffer, L.N., Molnar, R.J.: Appl. Phys. Lett. **78**(12), 1685 (2001). https://doi.org/10.1063/1.1356450
79. Anwar, A.F.M., Faraclas, E.W., Smith, K.V.: In: *Proceedings of the 2005 International Semiconductor Device Research Symposium*, pp. 221–222 (2005). https://doi.org/10.1109/ISDRS.2005.1596063
80. Kotani, J., Tajima, M., Kasai, S., Hashizume, T.: Appl. Phys. Lett. **91**(9), 093501 (2007). https://doi.org/10.1063/1.2775834
81. Chen, Y., Zhang, K., Cao, M., Zhao, S., Zhang, J., Ma, X., Hao, Y.: Appl. Phys. Lett. **104**(15), 153509 (2014). https://doi.org/10.1063/1.4871736
82. Goswami, A., Trew, R.J., Bilbro, G.L.: IEEE Trans. Electron Devices **61**(4), 1014 (2014). https://doi.org/10.1109/TED.2014.2302797
83. Rahemi, R., Li, D.: Scr. Mater. **99**, 41 (2015)
84. Albahrani, S.A., Mahajan, D., Kargarrazi, S., Schwantuschke, D., Gneiting, T., Senesky, D.G., Khandelwal, S.: IEEE Trans. Electron Devices **67**(2), 430 (2020)
85. Halford, D.: Proc. IEEE **56**(3), 251 (1968)
86. Hung, K.K., Ko, P.K., Hu, C., Cheng, Y.C.: IEEE Trans. Electron Devices **37**(5), 1323 (1990)
87. Dasgupta, A., Khandelwal, S., Chauhan, Y.S.: IEEE J. Electron Devices Soc. **2**(6), 174 (2014). https://doi.org/10.1109/JEDS.2014.2347991
88. Dasgupta, A., Khandelwal, S., Chauhan, Y.S.: IEEE Microwave Wireless Compon. Lett. **25**(6), 376 (2015). https://doi.org/10.1109/LMWC.2015.2422472
89. Wu, Y.F., Gritters, J., Shen, L., Smith, R., McKay, J., Barr, R., Birkhahn, R.: In: *The 1st IEEE Workshop on Wide Bandgap Power Devices and Applications*, pp. 6–10. IEEE, New York, (2013)
90. Parikh, P., Smith, K., Barr, R., Shono, K., McKay, J., Shen, L., Lal, R., Chowdhury, S., Yea, S., Smith, R., et al.: ECS Trans. **80**(7), 17 (2017)
91. Beach, R., Babakhani, A., Strittmatter, R.: Application note. In: *Efficient Power Conversion Corporation* (2011)
92. Shah, K., Shenai, K.: IEEE Trans. Electron Devices **59**(10), 2735 (2012)
93. Kotecha, R.M., Zhang, Y., Rashid, A., Zhu, N., Vrotsos, T., Mantooth, H.A.: In: *Proceedings of the 2017 IEEE Applied Power Electronics Conference and Exposition (APEC)*, pp. 2685–2691. IEEE, New York (2017)
94. Xiao, Y., Victory, J., Pearson, S., Sarkar, T., Challa, A., Dagan, M., Collanton, P., Andreev, C.: In: *Proceedings of the 2019 IEEE Applied Power Electronics Conference and Exposition (APEC)*, pp. 508–515. IEEE, New York (2019)
95. Khandelwal, S., Chauhan, Y.S., Fjeldly, T.A., Ghosh, S., Pampori, A., Mahajan, D., Dangi, R., Ahsan, S.A.: IEEE Trans. Electron Devices **66**(1), 80 (2018)
96. SI2: *Compact model coalition*. https://si2.org/cmc/
97. Khandelwal, S., Chauhan, Y.S., Hodges, J., Albahrani, S.A.: In: *Proceedings of the 2018 IEEE BiCMOS and Compound Semiconductor Integrated Circuits and Technology Symposium (BCICTS)*, pp. 93–97. IEEE, New York (2018)

98. Oh, S.Y., Ward, D.E., Dutton, R.W.: IEEE J. Solid-State Circuits **15**(4), 636 (1980)
99. Mahajan, D.D., Albahrani, S.A., Sodhi, R., Eguchi, T., Khandelwal, S.: IEEE Trans. Power Electron. **35**(6), 6332 (2020). https://doi.org/10.1109/TPEL.2019.2953060
100. McAndrew, C.C.: IEEE Trans. Electron Devices **53**(9), 2202 (2006)
101. Bendix, P., Rakers, P., Wagh, P., Lemaitre, L., Grabinski, W., McAndrew, C.C., Gu, X., Gildenblat, G.: In: *Proceedings of the IEEE 2004 Custom Integrated Circuits Conference (IEEE Cat. No. 04CH37571)*, pp. 9–12. IEEE, New York (2004)
102. Suárez, A.: IEEE Microwave Mag. **16**(5), 69 (2015). https://doi.org/10.1109/MMM.2015.2410951
103. Suárez, A., Ramírez, F., Jeon, S., Rutledge, D.: In: *Proceedings of the 2009 IEEE International Symposium on Radio-Frequency Integration Technology (RFIT 2009)*, pp. 203–206 (2009). https://doi.org/10.1109/RFIT.2009.5383668
104. Suárez, P.A.: pp. 340–343 (2017)
105. Collantes, J.M., Mori, L., Anakabe, A., Otegi, N., Lizarraga, I., Ayllon, N., Ramirez, F., Armengaud, V., Soubercaze-Pun, G.: IEEE Microwave Mag. **20**(7), 36 (2019). https://doi.org/10.1109/MMM.2019.2909516. https://ieeexplore.ieee.org/document/8726432/
106. Jugo, J., Portilla, J., Anakabe, A., Suarez, A., Collantes, J.M.: Electron. Lett. **37**(4), 226 (2001). https://doi.org/10.1049/el:20010170
107. Olivi, M., Seyfert, F., Marmorat, J.P.: Automatica **49**(2), 317 (2013). https://doi.org/10.1016/j.automatica.2012.10.005. http://www.sciencedirect.com/science/article/pii/S0005109812005122
108. Cooman, A., Seyfert, F., Olivi, M., Chevillard, S., Baratchart, L.: IEEE Trans. Microwave Theory Tech. **66**(1), 73 (2018). https://doi.org/10.1109/TMTT.2017.2749222
109. Chui, C.K., Chen, G.: *Signal Processing and Systems Theory*. Springer, Berlin (1992). https://doi.org/10.1007/978-3-642-88163-3
110. Martinez-Avendano, R.A., Rosenthal, P.: *An Introduction to Operators on the Hardy-Hilbert Space*. Springer, Berlin (2007)

Index

A
Access region charge density, 121
Access region resistance, 26, 28, 45
Accuracy, 16
AC symmetry, 148–149
Advanced SPICE Model for GaN HEMTs (ASM-HEMT)
 code, 173
 compact models (*see* Compact models)
 extraction flow process, 116
 formulation, 115
 GIT repository, 173
 model code download, 173
 model parameters, 52
 parameter extraction (*see* Parameter extraction of ASM-HEMT)
 Verilog-A code, 173
AlGaN barrier, 103, 110, 111, 113
Ambient temperature, 9–10
 compact models, 115
 device performance, 115
 electrical model parameters, 116
 extraction flow process, 116
 GaN HEMT (*see* GaN HEMT)
 gate current parameters, 122
 modeling of effect, 115
 self-heating effect and modeling, 122–124
 SPICE, 115
Analytical models, 84
ASM GaN, 18, 66, 102, 106, 132–134, 162, 165
ASM-HEMT, *see* Advanced SPICE Model for GaN HEMTs (ASM-HEMT)
ASM-HEMT compact model
 components
  additional model capability, 28–29
  core model, 27
  real device effects model, 27–29
 configurability, 23
 device geometry parameters, 21–23
 device physics parameters, 27
 features, 29–31
 model configuration switches, 23–26
 parameters, 21
 smoothing parameters, 27
Asymmetric time-constants, 65–66

B
Band-gap (E_g), 7
Barrier thickness dependence
 AM/PM, 166–169
Boost converter, 162, 163
Breakdown electric field, 7
BSIM, 15
Buck-boost converter, 161

C
Capacitance modeling, 140–142
Capacitances
 drain-voltage, 83
 field-plate, 83–90
 fringe, 84–85
 gate–drain, 83
 gate–source, 83
Carrier mobility, 119
Carrier saturation velocity, 92
Carrier scattering mechanisms, 166
CGSO, 84–85, 134, 142

S. Khandelwal, *Advanced SPICE Model for GaN HEMTs (ASM-HEMT)*,
https://doi.org/10.1007/978-3-030-77730-2

Channel-length modulation (CLM), 90, 92–95
  and drain–source voltage, 54
  I-V characteristics, 53
  output characteristics, 54
Charge-based formulation, 42
Charge-based GaN HEMT model, 18
Charge trapping, 96–98
Circuit model, 78
Circuit simulations, 9, 15
  GaN power device, 161–165
CLM, *see* Channel-length modulation (CLM)
Closed-loop frequency response, 152
CMC, *see* Compact model coalition (CMC)
CMC GaN Verilog-A model, 173
Compact model coalition (CMC), 17, 19, 137
  GaN, 137
  and industry standard GaN, 19
Compact models, 115
  ambient temperature, 9–10
  architecture/type, 16
  circuit simulations, 9
  CMC (*see* Compact model coalition (CMC))
  computational speed, 15
  electrical model of resistor, 9
  empirical, 11–12, 17
  GaN, 17–18
  global accuracy, 15
  industry standard GaN, 19
  Kirchhoff's current and voltage laws, 9
  measurement-based, 11, 17
  neural network based, 13–14
  physics-based, 14–15, 17
  predictability, 16
  properties, 15
  resistor, 10
  robustness, 16
  SPICE, 9
  TCAD, 12–13, 17
  terminal capacitances, 17
Computational speed, 15
Contact resistances, 121
Contact resistivity, 121
Core drain current model
  carrier movement, 43
  carrier transport, 43
  component, 45
  diffusion, 44
  Einstein relation, 44
  non-zero drain-source voltage, 44
  surface-potential model formulations, 42
Core drain–source current model, 50
Core model, 15
Coulomb scattering, 51
Current collapse, 2

D
DC model parameter extraction flow, 131–132
2-DEG, *see* Two-dimensional electron gas (2-DEG)
Design technology co-optimization
  non-linear AM/PM
    barrier thickness dependence, 166–169
    field-plate dependence, 169–172
Device geometry parameters
  ASM-HEMT model, 22
  device, 21
  features, 21
Device under test (DUT), 152
Diamond, 7
Diode current GaN, 101–102
Diode-like reverse current model, 101
Drain current
  FN model, 108–113
  formulation, 47, 151
  IDS, 125
  model, 166
  noise PSD, 125, 126
  PF model, 104–108
  reverse-bias (*see* Reverse-bias gate current)
  TE, 102–104
Drain-induced barrier lowering effect, 54–55
Drain-lag, 63, 96
Drain–source capacitance, 85, 89–90
Drain–source fringe field effect, 85
Drain–source symmetry, 146
Drain–source voltage, 170
DUT, *see* Device under test (DUT)
Dynamic losses, 6
Dynamic on-resistance, 96

E
EEHEMT, 18
EKV, 15
Electrical model of resistor, 9
Electrical model parameters, 116
Electrical parameters, 151
Electric field distribution, 90
Electric field profile, 90, 91
Electrode inductances, 133
Electron emission, 107
Electron mobility, 7
Empirical compact models, 11–12
Empirical GaN compact models, 18

F
Fast Fourier transform (FFT), 155
FDSOI, *see* Fully depleted silicon on insulator (FDSOI)
FFT, *see* Fast Fourier transform (FFT)
Field-plate capacitances
  configurations, 86
  drain–source capacitance, 89–90
  in GaN HEMTs, 85
  gate–drain capacitance, 87
  gate–source capacitance, 86–87
  source-or gate-electrode, 85
  2-DEG, 85–86
Field-plate dependence
  AM/PM, 169–172
Field-plates, 8, 17, 21–24
FinFET, 19
Flicker noise, 125–128
FN current, 108
Forward-bias condition, 122
Fourier coefficients, 155
Fowler–Nordheim (FN) model
  AlGaN barrier, 110, 111, 113
  characteristics, 109
  extracted parameters, 111, 113
  FN current, 108
  forward-bias and reverse-bias conditions, 111, 112
  gate-to-drain FN current source, 108
  gate-to-source FN current source, 108
  mechanism, 108
  PF model parameters, 111, 112
  Schottky barrier, 109, 110
  TE and PF currents, 108
  TE and PF models, 109
  TE model parameters, 109, 110
  vertical electric field, 111, 112
Frenkel–Poole current GaN, 102
Frequency response function (FRF), 153, 154
FRF, *see* Frequency response function (FRF)
Fringe capacitances, 84–85
Fully depleted silicon on insulator (FDSOI), 19

G
Gallium nitride (GaN)
  material properties (*see* GaN material properties)
  transistors (*see* GaN transistors)
GaN, *see* Gallium nitride (GaN)
GaN access region charge density, 121
GaN-based metal semiconductor FET (MESFET), 8
GaN charge trapping, 25
GaN C_{iss}, 140–142
GaN compact models
  empirical GaN compact models, 18
  field-plates, 17
  industry standard, 19
  measurement-based compact models, 18
  non-linear effects, 17
  and opportunities, 18
  physics-based GaN compact models, 18
  two-dimensional electron gas, 17
GaN contact resistances, 121
GaN C_{oss}, 140–142, 164, 165
GaN C_{rss}, 140–142
GaN cut-off voltage, 117–118
GaN device capacitance
  in ASM-HEMT model, 83
  and charge, 84
  charge trapping, 96–98
  cut-off voltage changes, 84
  drain-voltage, 83
  field-plates, 83–90
  fringe capacitances, 84–85
  gate–drain capacitance, 83
  gate-length, 84
  gate–source capacitance, 83
  high field region, 90–96
  multi-bias dependence, 83
  substrate bias effects, 98–100
  substrate capacitance, 98–100
  terminal charges, 84
GaN field plate effects, 151, 157, 161, 162, 164–166, 169–172
GaN gate current, *see* Gate current model
GaN HEMT, 8
  access region charge density, 121
  contact resistances, 121
  cut-off voltage, 117–118
  mobility, 118–119
  non-linearity, 167–172
  saturation velocity, 119–120
GaN material properties
  band-gap (E_g), 7
  breakdown electric field, 7
  electron mobility, 7
  key material parameters, 6
  power amplifier, 6
  power electronics application, 6
  saturation velocity, 7
GaN mobility, 118–119
GaN noise model, *see* Noise models
GaN physical model, 157–159
GaN physics, 27

GaN power device circuit simulations
  boost converter, 162, 163
  buck-boost converter, 161
  GaN transistor Q, 163
  OFF-state output capacitor, 163
  steady-state waveform, 163
  switching currents, 164
  transient simulation of boost converter, 163–165
GaN power device extraction flow, 137–146
GaN RF device extraction flow, 131–136
GaN saturation velocity, 119–120
GaN self-heating, 122–124
GaN SPICE model, 9, 115
GaN substrate effect, 98–100
GaN thermal resistance, 122–124
GaN transistor Q, 163
GaN transistors
  history, 7–8
  key technological advancement, 7–8
GaN velocity saturation, 40
Gate-and source-connected field-p, 8
Gate-connected field-plate, 88
Gate current model
  diode-based formulation, 101–102
  with drain current, 102–113
  Frenkel–Poole mechanism, 102
Gate current parameters, 122
Gate–drain capacitance, 85, 88
Gate–drain capacitance *vs.* gate voltage, 94, 96
Gate–drain fringe field effect, 85
Gate lag, 63, 96
Gate-length, 84
Gate resistance, 124
Gate–source capacitance, 83–87
Gate–source capacitance *vs.* gate voltage, 94, 95
Gate-to-drain current, 101
Gate-to-drain FN current source, 108
Gate-to-drain PF current source, 106
Gate-to-drain TE current, 103
Gate-to-drain voltage, 103
Gate-to-source current, 101
Gate-to-source FN current source, 108
Gate-to-source PF current source, 106
Gate-to-source TE current, 103
Gate-to-source voltage, 103
Gaussian distribution, 157
Geometry, 151
GITHUB repository, 173
Gummel symmetry test, 146–148

H
Hardy spaces, 154
Harmonic balance (HB) simulation, 148, 149, 152
Harmonic Transfer Function (HTF), 153
HEMT, *see* High electron mobility transistor (HEMT)
High electron mobility transistor (HEMT), 8
High field region on capacitance
  boundary conditions, 93
  carrier saturation velocity, 92
  CLM, 90, 92–95
  electric field distribution, 90
  electric field profile, 90, 91
  gate–drain capacitance *vs.* gate voltage, 94, 96
  gate–source capacitance *vs.* gate voltage, 94, 95
  Kirk effect, 93
  MOSFETs, 90
  quadratic formula, 93
  single-sided PN junction, 93
  TCAD, 90, 91, 94, 95
  2-DEG density, 92
Hilbert space, 154
HTF, *see* Harmonic Transfer Function (HTF)

I
Industry standard GaN compact models, 19
Intrinsic ASM GaN model, 133
I-V modeling, 60
I-V parameter extraction, 137–140

K
Kirchhoff's current and voltage laws, 9
Kirk effect, 93
Klaassen and Prins (KP) model, 128
Knee walkout, 63
Ku band, 158

L
Large signal GaN model, 79, 134–136, 152, 169
LEDs, *see* Light emitting diodes (LEDs)
Light emitting diodes (LEDs), 7
Load-pull simulations, 135, 136
Low-field carrier mobility, 119

M

Manufacturing variations in GaN
drain current formulation, 151
electrical parameters, 151
geometry, 151
I-V and C-V, 151
PA applications, 151
and RF device stability analysis
non-linear stability analysis, 156–161
python script, 153–156
Measurement-based and empirical compact models, 17
Measurement-based compact models, 11, 18
Measurement results
circuit model, 69
DC output, 67
I-V characteristics, 68
MESFET, *see* GaN-based metal semiconductor FET (MESFET)
Metal organic chemical vapor deposition (MOCVD), 8
Möbius transformation, 155
MOCVD, *see* Metal organic chemical vapor deposition (MOCVD)
Model configuration switches
architecture, 23
field-plate, 24
FPMOD combination, 24
FP1SMOD and FP1MOD, 25
GaN device, 25
GaN HEMTs geometry, 23
physical effects, 26
semiconductor, 26
Model extraction, 116, 121
Model quality tests
AC symmetry, 148–149
Gummel symmetry test, 146–148
harmonic balance simulations, 148, 149
reciprocity test, 149–150
Moore's law, 1
MOSFETs, 90
Multi-bias dependence of capacitance, 83
Multiple-input multiple-output (MIMO) systems, 152

N

Neural network based compact models, 13–14
Noise models
Flicker noise, 125–128
thermal noise, 127–130
Nominal device non-linear stability analysis
barrier thickness, 157, 159
electron mobility, 159
gate length, 157, 158
Gaussian distribution, 157
geometry and electrical parameters, 156
in Ku band, 158
measurement equation, 158
physical model parameter, 157, 158
pole-zero maps, 160
threshold voltage, 160, 161
of field-plate 1, 161
of field-plate 2, 162
2-DEG charge
distance of field-plate 1 to, 161
distance of field-plate 2 to, 162
Non-ideal sub-threshold slope
NFACTOR and CDSCD, 50, 51
Non-linear access region resistance model
contact resistance, 59
DIBL effect parameter, 56, 57
drain-side contact resistance, 61
fringe fields, 60
GaN HEMTs, 55
I-V behavior, 62
non-linear resistance, 60
saturation velocity, 58
source-side contact resistance, 61
2-DEG charge, 56
velocity, 57
Non-linear AM/PM
barrier thickness dependence, 166–169
behavior, 89
field-plate dependence, 169–172
Non-linear GaN model, 135–136
Non-linear large signal model, 135–136
Non-linear RF models, 151
Non-linear stability analysis
FRF, 153
HTF, 153
manufacturing variations, 156
methodology, 152
nominal device, 156–161
SRC3, 152–153
test-bed schematic, 152, 153
Normalized noise PSD, 127

O

OFF-state equivalent model, 142, 143
OFF-state output capacitor, 163
Ohmic contacts, 121
OIP3, 167, 168
Open-loop stability analysis, 152

P

Package parasitics
- components of power GaN device, 142, 143
- Im (z11), 144
- Im (z21/12), 144, 145
- Im (z22), 144, 145
- intrinsic FET device, 142
- OFF-state equivalent model, 142, 143
- Rem (z11), 144, 146
- RGC, 146
- RSC and RDC, 142
- Z-parameters, 144

Parameter extraction of ASM-HEMT
- GaN power device extraction flow, 137–146
- GaN RF device extraction flow, 131–136
- model quality tests, 146–150

Parasitic elements, 133
Parasitic networks, 134
Phase non-linearity, 168, 169
Physics-based compact models, 14–15, 17
Physics-based GaN compact models, 18
Physics-based modeling approach, 137
Piezo-electric polarization charge, 105
Pinch-off effect
- core model, 48
- surface-potential calculations, 47

Poole–Frankel (PF) reverse current, 102
Poole–Frenkel (PF) leakage, 102
Poole–Frenkel (PF) mechanism, 105
Poole–Frenkel (PF) model
- AlGaN barrier, 105, 106
- dislocations, 104
- electron emission, 107
- gate contact, 106
- gate leakage, 104
- gate-to-drain PF current source, 106
- gate-to-drain voltage, 106
- gate-to-source and gate-to-drain, 107
- gate-to-source PF current source, 106
- gate-to-source voltage, 106
- leakage current, 104
- mechanism, 105
- parameters, 111, 112
- piezo-electric polarization charge, 105
- surface-potential, 106
- trap-assisted tunneling mechanism, 107
- 2DEG, 106
- work-function difference, 107
- zero biasing condition, 107

Power amplifier application, 8
Power device extraction flow
- capacitance modeling, 140–142
- C-V, 137
- decoupled/independent formulations, 137
- I-V parameter extraction, 137–140
- package parasitics, 142–146
- physics-based modeling approach, 137
- silicon FETs, 137

Power electronics applications, 96
Power electronics switching application, 5–6, 8
Power spectral density (PSD), 125, 126, 128, 129
Power transistors, 6
Predictability, 16
PSD, *see* Power spectral density (PSD)
PSP, 15
Pulsed transfer characteristics, 73
Pulse measurement, 79
Python programming language, 152
Python script, 153–156, 158

Q

Quasi-Fermi levels, 71

R

Radio-frequency applications, 96
Radio-frequency (RF) power amplifier applications
- breakdown voltage, 3
- current and voltage swings, 2
- current-voltage plane, 3
- current-voltage relation, 3
- input power, 2
- load, 2
- load-line swing, 3
- matching networks, 2
- on-resistance, device, 3
- scaling g_m with width, 4, 5
- scaling I_{max} with width, 4, 5
- semiconductor device, 2
- semiconductor material properties, 4
- unity-gain frequency, 4

Real device effect models, 15, 27–28
Reciprocity test, 149–150
Research papers, 173
Resistor, 10
Reverse-bias conditions, 122
Reverse-bias gate current, 102
RF device stability analysis
- non-linear stability analysis, 156–161
- python script, 153–156

RF extraction
- DC model parameter extraction flow, 131–132

non-linear large signal model, 135–136
S-parameter model, 132–135
RF model, 29, 137, 151
RF modeling, 134
RF modeling methodology, 154
RGC, 146
Richardson's constant, 103, 109
Robustness, 16

S
Saturation velocity, 7, 119–120
Schottky barrier, 103, 109, 110
Schottky contact, 102
Schottky diodes, 103, 122
Self-consistent model, 98, 123
Self-heating, 122–124
Semiconducting technology, 6
Semiconductor, 118
SFP, *see* Source field-plate (SFP)
Shockley–Read–Hall (SRH), 71
Silicon-based transistors, 1
Silicon-on-insulator (SOI), 19
Silicon transistors, 51
Simulation program with integrated circuit emphasis (SPICE), 9, 115
Simulations, 115
design technology co-optimization, 166–172
GaN power device circuit simulations, 161–165
manufacturing variation and RF device stability analysis, 151–161
Single-input single-output (SISO) nature, 152
Single-sided PN junction, 93
Small-signal current source (SRC3), 152–153
Source-and drain-side contact resistances, 121
Source-connected field-plate, 88
Source field-plate (SFP), 169–172
S-parameter model, 132–135
SPICE models, 9
SRC3, *see* Small-signal current source (SRC3)
SRH-based trap model, 66
Static loss, 5, 6
Statistical models for GaN, 156, 157
Steady-state condition, 77–78
Substrate bias effects, 98–100
Substrate capacitance, 98–100
Surface-potential model
Fermi level, 37
regional expressions, 38
source-side, 38
Symmetric time-constants
modeling trapping effects, 64–65
trapping and detrapping, 64

T
Tbar, 168
TE, *see* Thermionic emission (TE)
Technical manual, 173
Technology computer-aided design (TCAD) models, 12–13, 17, 90, 91, 94, 95, 126–127
Terminal capacitances, 17
Terminal charge model
capacitance behavior, 41
capacitive current flow, 42
functions, 41
GaN HEMT device, 39
mid-point surface-potential, 40
SRH-based trap model, 66
Thermal capacitance, 122–124
Thermal dissipation, 8
Thermal model parameter, 116
Thermal network, 122–123
Thermal noise, 127–130
Thermal resistance, 122–124
Thermal simulations, 123
Thermal time-constant, 122
Thermionic emission (TE), 102–104, 109, 110
Thermometer, 124
Thick air bridge technology, 8
Trans-conductance, 130
Transistors
amplifying device, 1
application-level view
power electronics switching, 5–6
RF power amplifier, 2–5
deployed in electronics, 1
electronic switch, 1
semiconducting material, 1
silicon-based, 1
trans-conductance, 1
Trap-assisted tunneling mechanism, 107
Trap center, 74
circuit model, 76–77
control potential, 74–75
current injection, 75
Trapping models
AlGaN/GaN interface layer, 63
characteristics, 63
circuit implementation, 71

Trapping models (*cont.*)
  device, 63
  recombination-generation device, 64
  TRAPMOD, 63
Two-dimensional electron gas (2-DEG), 18, 85–86, 92, 106, 121
  analytical formulation models, 37
  carriers and device electrostatics, 34
  charge profile, 47
  compact model development, 36
  density, 34
  GaN HEMT device, 34
  gate voltage dependence, 33
  interpolation expression, 36
  source, 33
  symbols, 35

V
Velocity saturation effect
  drain–source voltage, 48
  drain voltage calculation, 49
  I-V characteristics, 49
  sub-threshold slope, 50

W
Wide band-gap semiconductor devices, 1
Work-function difference, 107

Z
Zero biasing condition, 107
Zone-based model, 18

MIX
Papier aus verantwortungsvollen Quellen
Paper from responsible sources
FSC® C105338

If you have any concerns about our products,
you can contact us on
ProductSafety@springernature.com

In case Publisher is established outside the EU,
the EU authorized representative is:
Springer Nature Customer Service Center GmbH
Europaplatz 3, 69115 Heidelberg, Germany

Printed by Libri Plureos GmbH
in Hamburg, Germany